It's another Quality Book from CGP

This book is for 11-14 year olds.

Don't be put off by our casual language.
Underneath the humour and chatty style we always make sure that our
books exactly match the requirements of the National Curriculum.

There are separate books for levels 3-6 and 5-8.

What CGP is all about

Our sole aim here at CGP is to produce the highest quality
books — carefully written, immaculately presented, and
dangerously close to being funny.

Then we work our socks off to get them out to you
— at the cheapest possible prices.

Contents

Big Numbers

You need to know how to:

> 1) *Read big numbers*   E.g., how would you say 1,432,678?
> 2) *Write them down*   E.g., how would you write *"Fourteen thousand, one hundred and ten"* as a number?

Groups of *Three*

It sounds sensible to me.

Always look at big numbers in *groups of three*.

4,521,396

So many MILLION   So many THOUSAND   And the rest
(i.e. 4 *million*, 521 *thousand*, 396) or written fully in words:
Four *million*, five hundred and twenty one *thousand*, three hundred and ninety six)

1) Always start from the extreme *right hand side* of the number →
2) Moving *left*, ← , put a comma in *every 3 digits* to break the number up
into *groups of 3*
3) Just *read each group of three* as a *separate number* and add "million"
and "thousand" on for the first two groups (going →, from the left).

Putting Numbers in *Order of Size*

Example: 49   220   13 3,402   76   94   105   684

Method

1) Put them into groups, the ones with fewest digits first:

> (all the 2-digit ones,   then all the 3-digit ones,   then all the 4-digit ones etc.)
> 49 13 76 94 | 220 105 684 | 3,402

2) For each separate group put them in order of size (easy):

> 13 49 76 94 | 105 220 684 | 3,402

The Acid Test

1) Write these numbers fully in words: a) 1,431,716   b) 25,999
   c) 6,812   d) 2,041   e) 1,801
2) Write this down as a number: Nine thousand, six hundred and fifty five.
3) Put these numbers in order of size: 102   4,600   8   59   26   3,785   261

Plus, Minus, Times and Divide

These are the building blocks — make sure you know how they work.

1) Plus and Minus are Opposites

THE STORY IN ENGLISH:

You start with 36p, someone gives you 64p, and you now have £1.00.

THE STORY IN MATHS:

$$36 + 64 = 100$$

Refusing to be bribed, you give back 64p and you're now left with what you started with, 36p.

$$100 - 64 = 36$$

2) Times and Divide are Opposites too

THE STORY IN ENGLISH:

You have 4 bags with 12 apples in each. You get an empty crate and pour them all in it, so the crate now contains 48 apples.

THE STORY IN MATHS:

$$12 \times 4 = 48$$

Changing your mind, you decide you want them back in their bags again, so you share the 48 apples among the 4 bags, ending with 12 apples in each, as before.

$$48 \div 4 = 12$$

3) Using Opposites when Checking

Example 1: *"What is the difference between 529 and 278?"*

Step 1) DO IT   529 - 278   on your calculator gives 251.

Step 2) CHECK IT   251 + 278   gets you back to 529.

Example 2: *"What is 342 ÷ 18?"*

Step 1) DO IT $342 \div 18 = 19$

Step 2) CHECK IT $19 \times 18 = 342$

The Acid Test

LEARN the methods on this page.

Do the following and then check they are correct by doing the opposite:

1) 27 + 49   3) 246 + 392   5) 14×5 7) 34×28

2) 65 - 36   4) 610 - 252   6) $100 \div 20$ 8) $240 \div 15$

Patterns with Times and Divide

You need to get the hang of this because they can ask you questions in the Exam based on this idea.

Example:

Suppose there's a crisis in the crisp industry and you find only half as many potato crisps in each bag as before. What do you do? Easy — to get the same number of crisps buy two bags instead of one. Or four instead of two. Or six instead of three. And so on. Doubling one number <u>makes up</u> for halving the other.

is the same as

Phew! Still 40 crisps in total!

Whichever way you look at it, it's still 40.

$$1 \times 40 = 40$$
$$2 \times 20 = 40$$
($\times 2$ ... $\div 2$)

There's an old saying,

"you gain on <u>the swings</u> what you lose on <u>the roundabout</u>".

Here you can call $\times$ "<u>the swings</u>" and $\div$ "<u>the roundabout</u>".

$$2 \times 20 = 40$$
$$4 \times 10 = 40$$
$$8 \times 5 = 40$$
($\times 2$ ... $\div 2$)

Here $\times 2$ regains what you lost with $\div 2$.

In the same way $\times 3$ makes up for $\div 3$.

$$5 \times 18 = 90$$
$$15 \times 6 = 90$$
$$45 \times 2 = 90$$
($\times 3$ ... $\div 3$)

The Acid Test

LEARN how these patterns work.

Find the missing number in each pattern:

1) $10 \times 2 = 20$
    $5 \times ... = 20$

2) $12 \times 6 = 72$
    $6 \times ... = 72$

3) $16 \times 3 = 48$
    $... \times 6 = 48$

4) $6 \times 15 = 90$
    $... \times 30 = 90$

5) $3 \times 26 = 78$
    $... \times 13 = 78$

6) $... \times 16 = 64$
    $8 \times 8 = 64$

Multiplying by 10, 100, 1000 etc.

You really should know this because

a) it's *very simple*, and   b) they're likely to *test you on it* in the Exam.

1) TO MULTIPLY ANY NUMBER BY 10

Move the Decimal Point ONE place BIGGER and if it's needed, ADD A ZERO on the end.

Examples:

$35.4 \times 10 = 354$

$162 \times 10 = 1620$

$8.625 \times 10 = 86.25$

2) TO MULTIPLY ANY NUMBER BY 100

Move the Decimal Point TWO places BIGGER and ADD ZEROS if necessary.

Examples:

$75.9 \times 100 = 7590$

$618 \times 100 = 61800$

$12.573 \times 100 = 1257.3$

3) TO MULTIPLY BY 1000, OR 10,000, the same rule applies:

Move the Decimal Point so many places BIGGER and ADD ZEROS if necessary.

Examples:

$28 \times 1000 = 28000$

$1.6789 \times 10,000 = 16789$

You always *move* the *DECIMAL POINT* this much:
1 place for 10,   2 places for 100,
3 places for 1000,   4 for 10,000   etc.

4) TO MULTIPLY BY NUMBERS LIKE 20, 300, 8000 ETC.

Multiply by 2 or 3 or 8 etc. FIRST,
then move the Decimal Point so many places BIGGER (➚)
according to how many noughts there are.

Wake me up when it's over.

Example:

To find 431×200,   *first multiply by 2* $431 \times 2 = 862$,
            then *move the DP 2 places* $= 86200$

The Acid Test

1) Work out   a) 14×100 b) 87.1×10 c) 25×1000
2) Work out   a) 3×200 b) 11×60 c) 7×3000

SECTION ONE — NUMBERS MOSTLY

Dividing by 10, 100, 1000 etc.

This is *pretty easy* too.   Just *make sure you know it* — that's all.

1) TO DIVIDE ANY NUMBER BY 10

Move the Decimal Point one place SMALLER and if it's needed, REMOVE ZEROS after the decimal point.

Examples:

$35.4 \div 10 = \underline{3.54}$

$162 \div 10 = \underline{16.2}$

$8.625 \div 10 = \underline{0.8625}$

2) TO DIVIDE ANY NUMBER BY 100

Move the Decimal Point 2 places SMALLER and REMOVE ZEROS after the decimal point.

Examples:

$75.9 \div 100 = \underline{0.759}$

$618 \div 100 = \underline{6.18}$

$12.573 \div 100 = \underline{0.12573}$

3) TO DIVIDE BY 1000, OR 10,000, the same rule applies:

Move the Decimal Point so many places SMALLER and REMOVE ZEROS after the decimal point.

Examples:

$528 \div 1000 = \underline{0.528}$

$16789 \div 10,000 = \underline{1.6789}$

You always *move* the *DECIMAL POINT* this much:
1 place for 10,   2 places for 100,
3 places for 1000,   4 for 10,000   etc.

4) DIVIDING BY 40, 300, 2000 ETC.

Hooray! I thought these two pages would never end.

DIVIDE BY 4 or 3 or 7 etc. FIRST and then move the Decimal Point so many places SMALLER (i.e. to the left).

Example:
To find $630 \div 300$,   *first divide by 3* $630 \div 3 = 210$,
              then *move the DP 2 places smaller* $= \underline{2.1}$

The Acid Test

1) Work out   a) $56 \div 10$ b) $426.5 \div 100$ c) $12.75 \div 1000$
2) Work out   a) $44 \div 20$ b) $666 \div 30$ c) $8000 \div 200$

6

Multiples and Factors

1) *Multiples are just "Times Tables"*

E.g. the *multiples of 2* are just the *2 times table*:

2   4   6   8   10   12   14   16   etc

The multiples of 8 are	8	16	24	32	40	48	etc
The multiples of 6 are	6	12	18	24	30	36	42	etc
The multiples of 12 are	12	24	36	48	60	72	84	etc

Finding Multiples with a Calculator

1) You can find the multiples of any number really easily *using your calculator*.
2) *Just keep adding the same number* — e.g. to find the multiples of 8 (The 8 times table) just press 8+8+8+8+ etc. and read the numbers off the display.

2) *Factors are just "The Numbers that Divide into Something"*

How to Find Factors

1) Use a calculator

2) Starting with 1, try all the numbers in turn, up to half the size of the number, to see if they divide.
   If they do, *they're factors*.

Example: Find the factors of 16.
Answer: *Using a calculator, divide 16 by each number in turn:*

16÷1 = 16   yes, so 1 *is a factor*
16÷2 = 8   yes, so 2 *is a factor*
16÷3 = 5.3   No, so 3 is NOT
16÷4 = 4   yes, so 4 *is a factor*
16÷5 = 3.2   No, so 5 is NOT
16÷6 = 2.6   No, so 6 is NOT
16÷7 = 2.29   No, so 7 is NOT
16÷8 = 2   yes, so 8 *is a factor*

This is now *halfway* (because 8 is half of 16) so we can STOP.

So the factors of 16 are 1, 2, 4, 8, and 16 itself don't forget.

The Acid Test

1) a) List all the multiples of 4 up to 100   b) List all the multiples of 9 up to 100 c) What is the first number that is a multiple of both 4 and 9?

2) a) Find all the factors of 6   b) Find all the factors of 15
   c) What number is a factor of both 6 and 15?

6

Odd, Even, Square & Cube Numbers

There are Four special sequences of numbers that you should KNOW:

1) EVEN NUMBERS

2 4 6 8 10 12 14 16 18 20 ...   i.e. the 2 times table

All *EVEN* numbers END with either a 0, 2, 4, 6 or 8   e.g. 144, 300, 612, 76

2) ODD NUMBERS

1 3 5 7 9 11 13 15 17 19 21 ...

All *ODD* numbers END with either a 1, 3, 5, 7 or 9   e.g. 105, 79, 213, 651

The EVEN numbers all Divide by 2   ODD numbers DON'T divide by 2

3) SQUARE NUMBERS

1	4	9	16	25	36	49	64	81	100	121	144 ...
(1x1)	(2x2)	(3x3)	(4x4)	(5x5)	(6x6)	(7x7)	(8x8)	(9x9)	(10x10)	(11x11)	(12x12) Etc.

They're called SQUARE NUMBERS because they're like the areas of this pattern of squares:

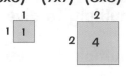 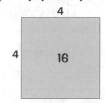

$1 \times 1 = \underline{1}$ $2 \times 2 = \underline{4}$ $3 \times 3 = \underline{9}$ $4 \times 4 = \underline{16}$

4) CUBE NUMBERS

1	8	27	64	125	216	343	512	729	1000...
(1x1x1)	(2x2x2)	(3x3x3)	(4x4x4)	(5x5x5)	(6x6x6)	(7x7x7)	(8x8x8)	(9x9x9)	(10x10x10)...

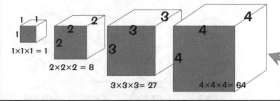

$1 \times 1 \times 1 = 1$
$2 \times 2 \times 2 = 8$
$3 \times 3 \times 3 = 27$
$4 \times 4 \times 4 = 64$

They're called CUBE NUMBERS because they're like the volumes of this pattern of cubes.

The Acid Test

1) Learn what EVEN and ODD NUMBERS are, and how to work out SQUARE NUMBERS and CUBE NUMBERS. Turn over and write down the first 10 of each.

2) From this list of numbers:

   27,   49,   100,   81,   125,   31,   132,   50

   write down   a) all the even numbers   b) all the odd numbers
                c) all the square numbers   d) all the cube numbers

Prime Numbers

Prime numbers can be _tricky_ — but they're a _lot less tricky_ if you just _learn_ these basics:

1) Basically, PRIME Numbers DON'T DIVIDE BY ANYTHING

And that's the best way to think of them.
So Prime Numbers are all the numbers that DON'T come up in Times Tables:

| 2 | 3 | 5 | 7 | 11 | 13 | 17 | 19 | 23 | 29 | 31 | 37 ... |

As you can see, they're an _awkward-looking bunch_ (that's because they don't divide by anything). For example:

The _only numbers_ that multiply to give 11 are 1×11

The _only numbers_ that multiply to give 23 are 1×23

In fact the _only way_ to get ANY PRIME NUMBER is $1 \times ITSELF$

2) They All End in 1, 3, 7 or 9

1) 1 is NOT a prime number

2) The first 4 primes are 2, 3, 5 and 7

3) 2 and 5 are the EXCEPTIONS because all the rest end in 1, 3, 7 or 9

4) But NOT ALL numbers ending in 1, 3, 7 or 9 are primes, as shown here:
(Only the circled ones are primes)

(2) (3) (5) (7)
(11) (13) (17) (19)
21 (23) 27 (29)
(31) 33 (37) 39
(41) (43) (47) 49
51 (53) 57 (59)
(61) 63 (67) 69

How to Find Prime Numbers

A Very Simple Method

1) <u>SINCE ALL PRIMES (above 5) END IN 1, 3, 7, OR 9</u>,

then to find a prime number between say, 70 and 80, <u>the only</u>

<u>possibilities</u> are: <u>71, 73, 77 and 79</u>

2) Now just <u>DIVIDE EACH ONE BY 3 AND 7</u> to find which of them

<u>ACTUALLY ARE PRIMES</u>. If it doesn't divide exactly by either 3 or 7

then it's a prime.

(This simple rule using just 3 and 7 is true for checking primes up to 120)

Example   "_Find all the primes between 70 and 80_"

1) The only possibilities are 71, 73, 77 and 79
2) So we now try to divide 71, 73, 77 and 79 by 3 and 7 to see
which of them actually are primes:

$71 \div 3 = 23.667$ so <u>71 IS a prime number</u>

$71 \div 7 = 10.143$ (_because it ends in 1, 3, 7 or 9_
       and it <u>doesn't divide by 3 or 7</u>)

$73 \div 3 = 24.333$ so <u>73 IS a prime number</u>

$73 \div 7 = 10.429$ (_because it ends in 1, 3, 7 or 9_
       and it <u>doesn't divide by 3 or 7</u>)

$79 \div 3 = 26.333$ so <u>79 IS a prime number</u>

$79 \div 7 = 11.286$ (_because it ends in 1, 3, 7 or 9_
       and it <u>doesn't divide by 3 or 7</u>)

$77 \div 3 = 25.667$

<u>BUT</u>: $77 \div 7 = 11$ so <u>77 is NOT a prime</u>, <u>BECAUSE IT</u>

11 is <u>a whole number</u>,   <u>WILL DIVIDE BY 7</u> $(7 \times 11 = 77)$

The Acid Test

LEARN the <u>3 Main Points on these pages</u>
<u>concerning Prime Numbers</u>.

Now <u>turn over and write down</u> what you have just learned.

1) Using the above method, find all the prime numbers between 80 and 100.

Ratio in The Home

There are lots of Exam questions which at first sight seem completely

different but in fact they can all be done using *The Golden Rule:*

DIVIDE FOR ONE, THEN TIMES FOR ALL

Example 1: *"5 loaves of Bread cost £2.70. How much will 3 loaves cost?"*

ANSWER: *The Golden Rule* says:

DIVIDE FOR ONE, THEN TIMES FOR ALL

which means:

> Divide the price by 5 to find how much FOR ONE LOAF, then
> multiply by 3 to find how much FOR 3 LOAVES.

So.....   £2.70 ÷ 5 = 0.54 = 54p   (for 1 loaf)

×3 = £1.62   (for 3 loaves)

Example 2: *"Divide £200 in the ratio 5:3"*

ANSWER: *The Golden Rule* says:

DIVIDE FOR ONE, THEN TIMES FOR ALL

The trick with this type of question is to add together the numbers in the RATIO to find
how many PARTS there are: 5 + 3 = 8 parts.   Now use The Golden Rule:

> Divide the £200 by 8 to find how much it is for ONE PART
>
> then multiply by 5 and by 3 to find how much 5 PARTS ARE
> and how much 3 PARTS ARE.

So.....   £200 ÷ 8 = £25   (for 1 part)

×5 = £125 (for 5 parts)   and ×3 = £75   (for 3 parts)

So £200 split in the ratio 5:3 is   £125 : £75

The Acid Test

1) If seven choc-bars cost 98p, how much will 4 choc-bars cost?
2) Divide £1260 in the ratio 5:7.

Rat 'n' Toad Pie

Example 3: The Recipe

The following recipe is for "Froggatt's Homespun Rat 'n' Toad Pie" and serves 4 people.

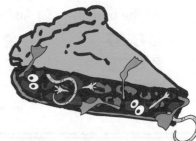

4 Freshly-caught Rats
2 Slimy Brown Toads
8 Ounces of "Froggatt's Hot Sickly Sauce"
12 Freshly-dug Potatoes
A big wodge of pastry

"Change these amounts so there's enough for TEN people."

ANSWER: *The Golden Rule* **says:**

DIVIDE FOR **ONE** , THEN TIMES FOR **ALL**

which means:

> DIVIDE each amount to get enough for <u>one person</u>,
> then <u>TIMES</u> to get enough for <u>TEN</u>.

Since the recipe is for *4 people* then <u>DIVIDE EACH AMOUNT BY 4</u> to find the amount for *1 person* — then <u>MULTIPLY THAT BY 10</u> to find how much for *10 people* — simple enough:

4 Rats ÷ 4 = <u>1 Rat</u> (for one person)   ×10 = <u>10 Rats</u> (For 10 people)

2 Toads ÷ 4 = <u>½ Toad</u> (for one person)   ×10 = <u>5 Toads</u> (For 10 people)

8 Ounces of "Froggatt's Hot Sickly
  Sauce" ÷ 4 = <u>2 Oz</u> (for one person)   ×10 = <u>20 Oz</u>   (For 10 people)

12 'taties ÷ 4   = <u>3 'taties</u> (per person)   ×10 = <u>30 'taties</u> (For 10 people)

A big wodge of pastry ÷ 4 then ×10 = A wodge of pastry <u>2½ times as big</u>.

In fact, <u>all the amounts</u> are just <u>2½ TIMES AS BIG</u>, if you notice.

The Acid Test

Work out the amount of each ingredient needed to
make enough Rat 'n' Toad pie for <u>8 people</u>.

Money

Questions about money give you good practice with decimals.

Adding *two sums of Money*

"What is £6.37 and £9.75?"

Remember to line up the decimal points.

1) Add the pence.

```
  £ 6.37
+ £ 9.75
———————
 ¹    2
```

2) Add the tens.

```
  £ 6.37
+ £ 9.75
———————
 ¹ ¹ .12
```

3) Add the pounds.

```
  £ 6.37
+ £ 9.75
———————
 ¹ ¹
 £16.12
```

So the answer is £16.12.

Subtracting *one sum of Money from Another*

"Ben has £15.65. He spends £10.99. How much money has he got left?"

1) Subtract the pence.

```
     ⁵ ₁
  £ 15.65
− £ 10.99
————————
        6
```

2) Subtract the tens.

```
   ⁴ ¹⁵ ₁
  £ 15.65
− £ 10.99
————————
      .66
```

3) Subtract the pounds.

```
   ⁴ ¹⁵ ₁
  £ 15.65
− £ 10.99
————————
  £  4.66
```

Line up the decimal points.

So the answer is £4.66.

Multiplying *Money*

"Neil buys 6 pots of paint at £4.50 each. What is the total cost?"

1) Multiply the pence.

```
  £ 4.50
×     6
———————
      0
```

2) Multiply the tens.

```
  £ 4.50
×   ₃ 6
———————
    .00
```

3) Multiply the pounds.

```
  £ 4.50
×   ₃ 6
———————
 £27.00
```

Therefore the cost is £27.00.

Dividing *Money*

"Jane has £4.76 to share equally between her four children. How much do they each get?"

1) Divide the pounds.

```
    1.
  ———————
4 | 4.76
```

2) Divide the tens.

```
    1.1
  ———————
4 | 4.7³6
```

3) Divide the pence.

```
    1.19
  ———————
4 | 4.7³6
```

Caution: begin at the pound end.

Each child receives £1.19.

The Acid Test

LEARN how to do money questions.

Now do these:

1) £4.92 + £2.65;

2) £20.50 – £4.05;

3) £6.99 × 3;

4) £18.30 ÷ 6.

The Best Buy

A favourite type of question they like to ask you in Exams is comparing the "value for money" of 2 or 3 similar items. Always follow *The Golden Rule:*

DIVIDE BY THE PRICE, *IN PENCE*

(*TO GET THE AMOUNT* *PER PENNY*)

Example The local Royalty Cinema sells popcorn in three different sizes, Small, Regular and Family Size. The question is: Which of these represents "THE BEST VALUE FOR MONEY"?

400g at £2.62   250g at £1.90   100g at 89p

ANSWER: *The Golden Rule* says:

DIVIDE BY THE PRICE *IN PENCE* *TO GET THE AMOUNT* *PER PENNY*

So we shall:

$$400g \div 262p = \text{1.5g PER PENNY}$$
$$250g \div 190p = \text{1.3g PER PENNY}$$
$$100g \div 89p = \text{1.1g PER PENNY}$$

So we can now see straight away that THE 400g BOX IS THE BEST VALUE FOR MONEY because you get MORE POPCORN PER PENNY

(As you should expect, it being the big box).

With any question comparing "value for money", *DIVIDE BY THE PRICE* (in pence) and it will always be THE BIGGEST ANSWER IS THE BEST VALUE FOR MONEY.

The Acid Test

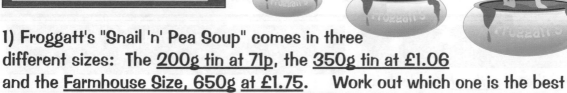

1) Froggatt's "Snail 'n' Pea Soup" comes in three different sizes: The 200g tin at 71p, the 350g tin at £1.06 and the Farmhouse Size, 650g at £1.75. Work out which one is the best value for money. (And don't just guess!)

Times and Divide without a Calculator

There will be some questions in the Exam which you won't be allowed to do on your calculator. Practice makes perfect...

Long *Multiplication*

This is easy when you realise it's just a few *short* multiplications added together.

Example | "What is 46×14?"

THIS IS EASY: 46 × 14 is just 46×4 and 46×10.

1) First we work out 46×4, which can also be split into two parts.

$$4×6 \quad \begin{array}{r} 4\,6 \\ \times\ 1\,4 \\ \hline \,2 \\ 4 \end{array} \quad \text{and} \quad 4×4 \quad \begin{array}{r} 4\,6 \\ \times\ 1\,4 \\ \hline 1\,8\,4 \end{array}$$

2) Add 0 to the next line (to prepare for multiplying by a number in the 1 0 's column)

$$\begin{array}{r} 4\,6 \\ \times\ 1\,4 \\ \hline 1\,8\,4 \\ 0 \end{array}$$

3) Now work out 46×10.

$$1×6 \quad \begin{array}{r} 4\,6 \\ \times\ 1\,4 \\ \hline 1\,8\,4 \\ 6\,0 \end{array} \qquad 1×4 \quad \begin{array}{r} 4\,6 \\ \times\ 1\,4 \\ \hline 1\,8\,4 \\ 4\,6\,0 \end{array}$$

4) Finally, add: 184
to 460
to get 644

Division | There are *two* options:

EITHER **1) LEARN HOW TO DO IT PROPERLY**

Example | "What is 864 ÷ 8?"

$$\begin{array}{r} 1 \\ 8\,|\overline{8\,6\,4} \end{array} \qquad \begin{array}{r} 1\,0 \\ 8\,|\overline{8\,6^6\,4} \end{array} \qquad \begin{array}{r} 1\,0\,8 \\ 8\,|\overline{8\,6^6\,4} \end{array}$$

8 into 8 goes once   8 into 6 won't go so carry 6   8 into 64 goes 8 times

OR **2) TAKE A GUESS AND USE MULTIPLICATION TO CHECK**

(See P.2)

Example | "What is 92 ÷ 4?"

Try 20   4 × 20 = 80   = too small

Try 25   4 × 25 = 100   = too big

Try 23   4 × 23 = 92   SPOT ON.   So 92 ÷ 4 = 23

The Acid Test | LEARN the above methods for *times* and *divide*.

Try these *without* a calculator:

1) 28 × 12;   2) 56 × 11;   3) 104 × 8;

4) 96 ÷ 8;   5) 242 ÷ 2;   6) 84 ÷ 7.

Typical Questions with × and ÷

Example 1

"Flora's Florists sell begonias at £2.50 each. How much would 16 begonias cost?"

16 × £2.50 = 8 × £5.00
= <u>£40</u>

Example 2

"Tulips sell at £1.95 a bunch. How much would 4 bunches cost?"

1 bunch   = £2   less 5p
So 4 bunches = 4×£2   less 4×5p
                      = £8   less 20p
                      = <u>£7.80</u>

Example 3

"How many tulips could you buy with £12?"

£2 buys 1 bunch with 5p change.

So 6 × £2 buys 6 bunches with 6×5p change,

Which means £12 buys 6 bunches with 30p change.

Since the 30p change is not enough to buy another bunch, <u>6</u> is the most you'll get.

Example 4

"What would be the cost of 2 begonias and 2 bunches of tulips?"

2 begonias and 2 bunches of tulips cost:

2×£2.50 + 2×£1.95

which is the same as   £5 + 2×£2 less 2×5p

= £5 + £4 less 10p
= <u>£8-90</u>

The Acid Test

Do these question *without* a calculator.

1) Posh Pencils cost £1.49 each. How much would five pencils cost?

2) Crumby Crackers cost £1.10 a packet. How many packets can you buy for £12?

Calculator Buttons 1

There are few things in life quite as grim as watching someone hacking away at their calculator, constantly bashing the overworked cancel button every few seconds, *and never getting any better at it.* Here are some nice tricks to save you a lot of button-bashing.

1)

1) **Press the buttons** <u>slowly</u> **and** <u>carefully</u>
— one small mistake can lose you a lot of marks.

2) <u>Watch</u> **the display as you press** <u>each button</u>
to make sure you have actually pressed it — many mistakes result from keys not being pressed properly.

3) <u>Press</u> **=** <u>at</u> **the end** <u>of every</u> **calculation**
People often get the wrong answer just because they forget to press **=** at the end.

2) **C** *(SEMI-CANCEL)* **and** **AC** *(ALL CANCEL)*

(Alternatively, with **on/c** or **CE/C** press <u>ONCE</u> for *semi-cancel* and <u>TWICE</u> for *all cancel*)

Please, for goodness' sake, you must kick the habit of swiping the **AC** button every time things go slightly wrong. Semi-cancel is MUCH better if you know what it does:

IT ONLY CANCELS THE NUMBER YOU ARE ENTERING.

Everything else remains intact. If you learn to use **C** instead of **AC** for correcting wrong numbers you'll <u>HALVE</u> the time you spend pressing calculator buttons!

3) **The** <u>SELF-CANCELLING</u>

<u>function</u> **buttons:** **+** **−** **×** **÷**

The thing to remember here is that these four buttons are all <u>SELF-CANCELLING</u>, which means if you press **+** and then **÷**, your calculator will totally ignore the **+** and just do the **÷** instead.

<u>So</u>: *If you press the wrong function button, just <u>ignore it</u>, <u>press the right one</u>,* and <u>*CARRY ON*</u>. Try **9** **+** **÷** **×** **−** **6** **=** to see how well it works.

4) **The** <u>PLUS/MINUS</u> **button** **+/−**

All this button does is reverse the + or − sign <u>OF THE NUMBER ALREADY IN THE DISPLAY</u>. Its main use is for *<u>entering negative numbers</u>*.
For example to work out -6 × -2 you'd press

6 **+/−** **×** **2** **+/−** **=** (Which gives 12 (not −12))

Note that you only press **+/−** <u>*AFTER you've entered the number*</u>.

Calculator Buttons 2

5) The SECOND FUNCTION Button [SHIFT] (or [2nd] or [INV])

Most calculator buttons do _2 functions_. The main function is written on the button itself, whilst the 2nd function is written _above it_.

To use the 2nd function of any button just press [SHIFT] first.
(or [2nd] or [INV] if that's what your calculator has)

(If you have a fancier calculator many buttons may perform _3 functions_. Fortunately they are colour-coded so the colour on the [SHIFT] (or [2nd] or [INV]) button is matched by the colour of 2nd function written above other buttons.)

6) Two tricks for Reading Displayed Answers

1) Putting "× 10" in for Numbers that are _too big_!

Sometimes your answer will be too big for the calculator display and instead you'll get two extra little numbers up in the air at the end of the display like this:

$$7.532^{11}$$

To give this as a proper answer _you have to remember to put "× 10" into your answer_ like this: 7.532×10^{11}. So don't forget!

2) Remembering that your answer is in £ and Pence

The other trick has to do with answers that are in £ and pence.
You have to think carefully what the display actually means.
Look at these examples:

A display showing 2.5 means the answer is _£2.50_
A display showing 0.37 means the answer is _37p_
A display showing 8.63621 means the answer is _£8.64_

The Acid Test

1) What are the two different types of cancel? Explain the difference between them.

2) What does the [SHIFT] button do? When would you need to use it?

3) What would you press to work out -5 × -3?

4) How should you write this on paper: 9.16^{14}

5) How many £ and pence is this likely to be: 3.0964

Using Formulas

THESE AREN'T AS BAD AS YOU THINK so long as you follow the step by step method and do things in BODMAS order.

Step by Step Method

1) WRITE OUT THE FORMULA

2) WRITE IT AGAIN DIRECTLY UNDERNEATH but this time with *numbers in place of all the letters*.

3) WORK IT OUT IN STAGES
Use BODMAS and *write down values for each bit* as you go along.

4) DON'T TRY AND DO IT ALL IN ONE GO ON YOUR CALCULATOR — a *ridiculous method* that fails *at least 50% of the time*.

Example:
If P = 2(L + w), find P when L = 2 and w = 5

Step 1)   P = 2(L + w)

Step 2)   = 2 (2 + 5)

Step 3)   = 2 (7)
= 2×7
= 14

BODMAS

The word BODMAS tells you the right order to work things out in any formula.   The letters stand for:

Brackets (take priority) **O**ver **D**ivision, **M**ultiplication, **A**ddition, **S**ubtraction

and what it means is:
Work out anything in brackets first, then any numbers that need *dividing*, then any that need *multiplying*, then do *add-ups*, and then *take-aways*.

The Tricky one

"Work out the value of $\frac{12 + 48}{15 \times 3}$ "

It's no good just pressing [12] [+] [48] [÷] [15] [×] [3] [=] — it will be *completely wrong*.

The calculator will think you mean $12 + \frac{48}{15} \times 3$ because the calculator will do the *division and multiplication* BEFORE it does the *addition* (BODMAS).

YOU MUST WORK IT OUT IN STAGES, top and bottom first, write them down, and *only then* divide them, like this:
$$12 + 48 = \underline{60}, \quad 15 \times 3 = \underline{45} \quad \Rightarrow \quad 60 \div 45 = \underline{1.333}...$$

The Acid Test

LEARN the 4 steps of the Strict Method and exactly what BODMAS stands for.

1) Turn over the page and write down what BODMAS stands for.
2) If A = F(2 + H) and F = 3 and H = 4, find the value of A.
3) If R = 2P + 3Q, find R when P = -5, Q = 2.

SECTION ONE — NUMBERS MOSTLY

Ordering Decimals

Arrange in _Order of Size_

In your Exam you will be expected to _arrange_ a list of decimal numbers in _order of size_.

This means, for example, knowing that 0.305 is _greater_ than 0.0799.

This is easy when you place the numbers underneath each other and line up the decimal points.

It can be seen that <u>0.3050</u> is _more_ than <u>0.0799</u>	Just as <u>3050</u> is _more_ than <u>799</u>

Unfortunately things can get a bit trickier, especially if the list is long and the numbers are _not placed_ tidily _underneath_ one another. Also their lengths _vary_ a great deal. So when in doubt, go for:

The Foolproof _Method of Ordering Decimals_

Example:
"Arrange the following in increasing order of size:

0.708   1.020   0.215   0.00987   0.03006".

Step 1:

Arrange them in a column with the _decimal points underneath one another_:

```
0.708
1.020
0.215
0.00987
0.03006
```

Step 2:

Make them all the _same length_ by _filling in extra zeroes_:

```
0.70800
1.02000
0.21500
0.00987
0.03006
```

Step 3:

Ignore the decimal points and just treat the numbers as _whole numbers_:

```
70800
102000
21500
987
3006
```

Step 4:

Arrange them in order:

```
987
3006
21500
70800
102000
```

Last Step:

Put the _decimal points_ and _beginning zeroes back in_:

```
0.00987
0.03006
0.215
0.708
1.020
```

The Acid Test

LEARN the <u>5-step method</u> on this page.

Then use it to order the following list:

1.03, 0.792,   0.0591,   0.006,   0.082,   0.00049.

Revision Test for Section One

WHAT YOU'RE SUPPOSED TO DO HERE is put all the methods of Section One into practice to answer these questions.

Revision Test

1) Write this number out *in words*: 4,216,386

2) Put these numbers *in order of size*:

      144   26   1,212   4   48   612   842   2006

3) *Without* using your calculator, work out:
        a) 26.8×100
        b) 340×1000
        c) $648 \div 1000$
        d) 30×20
        e) $4000 \div 20$

4) What are *Multiples?* List the *first six multiples* of 10 and also of 3.

5) What are *Factors?* Find *all the factors* of 24.

6) From this list: 9, 18, 1, 25, 63, 100, 36, 16

                pick out   a) The *square numbers*
                               b) The *even numbers*
                               c) The *odd numbers*

7) What are *EVEN NUMBERS?* Write down *the first ten* of them.

8) What are *ODD NUMBERS?* Write down *the first ten* of them.

9) a) What are *SQUARE NUMBERS?* Write down *the first twelve* of them.
   b) What are *CUBE NUMBERS?* Write down *the first ten* of them.

10) What are *PRIME NUMBERS?* Write down *the first six* of them.

11) What *rules* are there for finding *prime numbers* (less than 120)?
    Work out *which of the following* are prime numbers:

        5, 0, 11, 23, 7, 29, 14, 19, 15, 17, 13

Revision Test for Section One

12) What is _The Golden Rule_ for <u>Ratio In the Home</u>?

13) If 7 pints of milk cost £2.66, _how much will 5 pints cost_?

14) If 9 packs of orange jellies weigh 675g, _how much will 4 packs of jellies weigh_? (Remember: "Divide for one, then times for all")

15) Work out _without_ your calculator:
   a) £4.75 + £2.50;   b) 6 × £7.99;   c) 10 × £12.25.

16) What is _The Golden Rule_ for finding the _"Best Buy"_?
   Two different sized pineapples are on sale in a shop:
   Which one do you _expect_ to be _the best value_?
   Do a _quick calculation_ to work out _for sure_
      which one is the "Best Buy".

17) _Without your calculator_, do 43 × 28, then divide the answer by 28 and check you get back to 43.

18) What are the two different _cancel buttons_ on your calculator?

19) What does _each type_ of cancel do?

20) What should you do if you press ➕ _by mistake_, instead of ➗ ?

21) Which button is the _2nd function button_? What does it do?

22) How would you _enter_ the number _-6_ into the calculator?

23) _Work out_ -6 × -5 with your calculator.

24) How would you _write this down_ as an answer: `2.4` ^{11}

25) What is this answer _in £ and pence_: `5.324`

26) What are the _FOUR Steps_ for working out formulas?

27) If P = 2(a + b), _find P_ when a = 1 and b = 6.

28) Work out the value of $\frac{13+62}{40-15}$ ... _and don't be hasty_.

29) _Arrange_ in increasing order 0.5; 0.51; 0.15; 0.05; 0.55; 0.505.

SO IF YOU GET STUCK with any of these questions, _have a look back at the right page_ in Section One to _find out how to do it_!

Perimeters

Perimeter is the distance *all the way around the outside of a 2D shape*.

Finding the Perimeters of shapes

To find a PERIMETER, you ADD UP THE LENGTHS OF ALL THE SIDES, but....THE ONLY RELIABLE WAY to make sure you get *all the sides* is this:

1) PUT A BIG BLOB AT ONE CORNER and then go around the shape

2) WRITE DOWN THE LENGTH OF EVERY SIDE AS YOU GO ALONG

3) EVEN SIDES THAT SEEM TO HAVE NO LENGTH GIVEN
  — you must *work them out*

4) Keep going until you get back to the BIG BLOB.

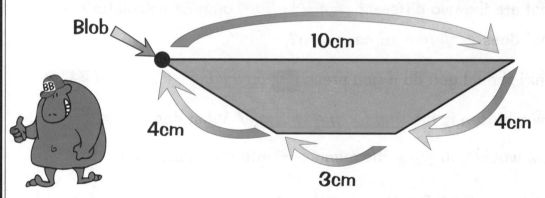

Blob

10cm

4cm

4cm

3cm

E.g. 10 + 4 + 3 + 4 = <u>21cm</u>

Yes, I know you think it's *yet another fussy method*, but believe me, it's so easy to miss a side. <u>You must use good RELIABLE METHODS for EVERYTHING</u> — or you'll lose marks willy nilly.

The Acid Test

LEARN THE RULES for finding
perimeters.

7cm

3cm

6cm

4cm

1) <u>Turn over and write down</u> what you have learnt.

2) Find the perimeter of the shape shown here:

Areas

They don't promise to give you these formulae in the Exam, so if you don't learn them before you get there, you'll be SCUPPERED — simple as that.

1) *RECTANGLE*

The area of a rectangle is *easy* if you know your times tables.

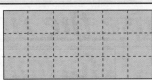

There are 3 lots of 6 squares in this rectangle. So the area is just $3 \times 6 = 18$.

In the exam, the shape might not have squares in it, so

LEARN THIS FORMULA:

Area of rectangle = Length × Width

$$A = L \times W$$

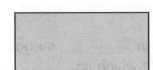

Width

Length

2) *TRIANGLE*

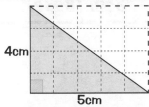

4cm

5cm

Area of rectangle
$= 5 \times 4 = 20 \text{cm}^2$

Counting squares isn't as good for triangles because the squares aren't all complete. However, *right-angled triangles*, like the one shown, are easy. The area is half that of the rectangle. For this one:—

$$\frac{1}{2} \times 5 \times 4 = 10 \text{cm}^2$$

"Height" and "Base" are the words used for triangles.

LEARN THIS FORMULA:

Area of triangle = ½ × Base x Vertical Height

$$A = \frac{1}{2} \times B \times H_v$$

Height

Base

A right-angled triangle.

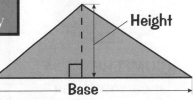

Height

Base

For other triangles, the *height* must always be the *vertical height*, never the sloping height.

The Acid Test

LEARN THE WHOLE OF THIS PAGE

Then turn over and write down as much as you can from memory. Then try again.

Areas

KNOWING THE FORMULAS on the previous page is *an important first step*, but here are a few other things you'll need to know IF YOU ACTUALLY WANT TO GET THE QUESTIONS RIGHT:

Identify *the shape and use the* right *formula*

A lot of people lunge straight in with the first formula to come into their head and *never imagine they might have the wrong formula.*

This is a TRIANGLE so you *can't use the rectangle formula* which is: AREA = LENGTH × WIDTH

You must use the *triangle formula*:

"AREA = ½ × BASE × HEIGHT"

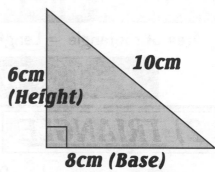

10cm

6cm (Height)

8cm (Base)

CIRCLES

There are *two formulas* for *circles*.

DON'T MUDDLE THEM UP!

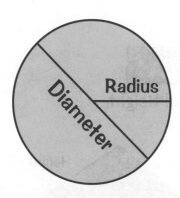

Radius

Diameter

CIRCUMFERENCE = distance round the outside of the circle

1) AREA of circle = π × (radius)²

$$A = \pi \times r^2$$

E.g. if the radius is 4cm
A = 3.14 x (4x4)
= 50cm²

2) CIRCUMFERENCE = π x Diameter

$$C = \pi \times D$$

E.g. if the radius is 4cm
Then diameter = 8cm,
C = 3.14 x 8
= 25.12cm

The Acid Test

LEARN THE CIRCLE FORMULAS.

Then cover the page and write them out. Then try again.

Areas — Typical Questions

Don't reach straight for the calculator

You might be _kidding yourself_ that it _"takes too long"_ to write down your working out — but what's so great about getting ZERO MARKS _for an easy question_?
Compare these two answers for finding the area of the triangle opposite:

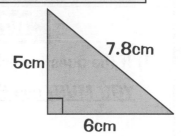

7.8cm
5cm
6cm

ANSWER 1:   32   <u>30</u>   X

ANSWER 1 gets <u>NO MARKS AT ALL</u> — 30 is the wrong answer and there's nothing else to give any marks for.

ANSWER 2:   A = ½ × B × H   ✓
           = ½ × 6 × 5   ✓
           = <u>15 cm²</u>   ✓

ANSWER 2 has _3 bits that all get marks_, — so even if the answer was wrong it would still get most of the marks!

The thing is though, when you _write it down step by step_, you can see what you're doing _and you won't get it wrong in the first place_   —   try it next time, go on .... just for the wild experience.

Show _your_ working out

1) <u>Write down the formula or method</u>:   A = ½ × B × H   ✓
2) <u>Write the numbers in</u>, on the paper: = ½ × 6 × 5   ✓
3) THEN <u>work it out carefully</u> with your calculator: = <u>15 cm²</u>   ✓

Remember, you basically get one mark for each step, as in ANSWER 2 above ...
If you decide you're TOO CLEVER _or_ TOO BUSY to bother with showing your working out, then good luck to you — you'll certainly need it.

The Acid Test

MEMORISE THE RULES FOR FINDING THE AREA OF A SHAPE.

Then find the areas of these 2 shapes — and make sure you follow the rules you've just learnt!

1)
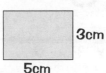
3cm
5cm

2)
4m
6m

3) Area = 18m²
Find the missing side.
6cm
?

4) Area = 24cm²
Find the missing side.
3cm
?

Circle Questions

The Big Decision:

"Which circle formula do I use?"

1) If the question asks for "the area of the circle",

   YOU MUST use the FORMULA FOR AREA:

   $$A = \pi \times r^2$$

2) If the question asks for "circumference" (the distance around the circle)

   YOU MUST use the FORMULA FOR CIRCUMFERENCE:

   $$C = \pi \times D$$

AND REMEMBER, it makes *no difference at all* whether the question gives you *the radius* or *the diameter*, because whichever one they give you, it's <u>DEAD EASY</u> to work out the other one — the diameter is always <u>DOUBLE</u> the radius.

Example 1:

"Find the circumference and the area of the circle shown below."

<u>ANSWER</u>:

The radius=5cm,   so the <u>Diameter=10cm</u>   (easy huh?)

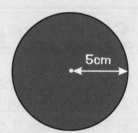
5cm

Formula for *circumference* is:

$C = \pi \times D$, so
$C = 3.14 \times 10$
   $= \underline{31.4cm}$

Formula for <u>AREA</u> is:

$A = \pi \times r^2$
   $= 3.14 \times (5\times5)$
   $= 3.14 \times 25 = \underline{78.5cm^2}$

Example 2: The good old "Bicycle Wheel" question:

This is a very common Exam question.
"How many turns must a wheel of diameter 1.1m make to go a distance of 20m?"

<u>ANSWER</u>:

Each full turn moves it *one full circumference* across the ground, so

1) find the circumference using "$C = \pi \times D$" : $C = 3.14 \times 1.1 = \underline{3.454m}$

2) then find how many times it fits into the distance travelled, by dividing:
     i.e. $20m \div 3.45m = 5.79$ so the answer is <u>5.8 turns of the wheel</u>.

Extra Circle Details

1) π *"A Number a Bit Bigger than 3"*

The big thing to remember is that π (called "pi") only seems confusing because it's a scary-looking Greek letter.   In the end, it's just an *ordinary number* (3.14159...) which is rounded off to either 3 or 3.14

(depending on how accurate you want to be).

And that's all it is: A NUMBER A BIT BIGGER THAN 3

2) *Diameter is TWICE the Radius*

The DIAMETER goes *right across* the circle

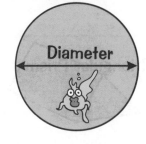
Diameter

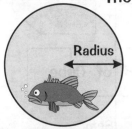
Radius

The RADIUS only goes *halfway* across

Examples:
If the radius is 3cm, the diameter is 6cm,
If the radius is 10m, the diameter is 20m
If D = 8cm, then r = 4cm,
If diameter = 4mm, then radius = 2mm.

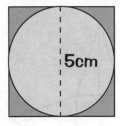
Remember: the DIAMETER IS EXACTLY DOUBLE THE RADIUS

Example:

"A circle with D=5cm is drawn inside a square so that they touch. Find the area of the square and the area of the circle".

5cm

First find the area of the circle:
$$A = \pi r^2$$
$$= 3.14 \times (2.5 \times 2.5) \qquad \text{since } r = \tfrac{1}{2}D$$
$$= 19.625$$
Area of circle = 19.6cm².

The square must have sides of length 5cm (look at the diagram to see if you agree). So,   Area of square = 5 × 5
$$= 25cm^2.$$

As you'd expect, the area of the square is a bit more than the area of the circle.

The Acid Test

LEARN ALL THE MAIN POINTS on these two pages. They are all *mighty important*.

Then turn over and see if you can write them down.

1) A circular table has a radius of 25cm. Find its area and circumference using the methods you have just learnt. Remember to show all your working out.

2) How many turns will a hoop of diameter 1m make if it rolls 628m?

Solids and Nets

You need to know what *Face*, *Edge* and *Vertex* mean:

Vertex (corner)

Face

Edge

A NET is just
A SOLID SHAPE FOLDED OUT FLAT
These are the 4 that you need to know really well for the Exam:

1) Triangular Prism

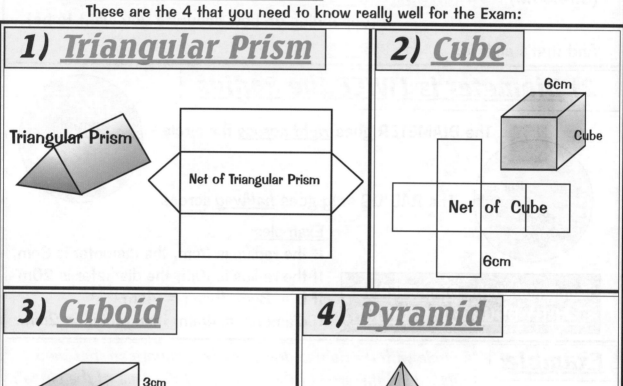

Triangular Prism

Net of Triangular Prism

2) Cube

6cm

Cube

Net of Cube

6cm

3) Cuboid

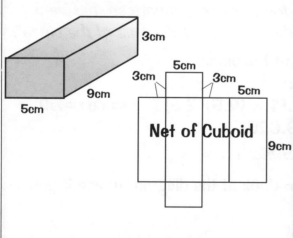

3cm

5cm

3cm   3cm

9cm   5cm

5cm

Net of Cuboid

9cm

4) Pyramid

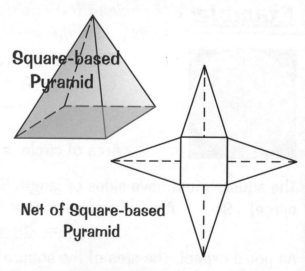

Square-based
Pyramid

Net of Square-based
Pyramid

Working out the Area of a Net

They might well ask you to draw one of the nets on this page. They might also ask you to work out the area of it. This is what you do;

1) Work out the areas of all the SEPARATE RECTANGLES and TRIANGLES.
2) Then ADD THEM ALL TOGETHER. Try it for the cube above.

SECTION TWO — SHAPES

Volume and Capacity

VOLUMES — YOU MUST LEARN THESE TOO!

1) *CUBE*

Volume of Cube = Side × Side × Side
= (Side)³

(The other word for volume is <u>CAPACITY</u>)

S
S
S

$$V = S \times S \times S$$

or

$$V = S^3$$

2) *CUBOID* *(Rectangular Block)*

Volume of Cuboid = Length × Width × Height

Height

$$V = L \times W \times H$$

Width
Length

The Acid Test

LEARN THE <u>2 volume formulas</u>, then cover them up and write them down from memory.

1) How many small cubes are there in this cuboid?

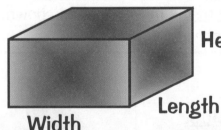

2) What is the volume of this box?

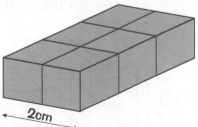

2cm

3) Find the volume of a large cube which has sides measuring 2m.

4) A box measures 8cm×4cm×3cm. Draw a picture of it and find its volume. Try drawing its net.

5) Cover up the opposite page and draw the 4 solids and their nets.

Symmetry

SYMMETRY is where a shape or picture can be put in DIFFERENT POSITIONS that LOOK EXACTLY THE SAME. There are THREE TYPES of symmetry:

1) _Line Symmetry_

This is where you can draw a MIRROR LINE (or more than one) across a picture and _both sides will fold exactly together_.

| 1 LINE OF SYMMETRY | 2 LINES OF SYMMETRY | NO LINES OF SYMMETRY | 1 LINE OF SYMMETRY | 3 LINES OF SYMMETRY |

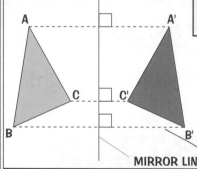

How to draw a reflection:

1) Reflect each point one by one

2) Use <u>a line which crosses the mirror line at 90° and goes EXACTLY the same distance on the other side of the mirror line</u>, as shown.

MIRROR LINE   A line which crosses at 90° is called _a perpendicular_

2) _Plane Symmetry_

No, not that kind of plane...

Plane Symmetry is all to do with 3-D SOLIDS.

Just like _flat shapes_ can have a _mirror line_, so _solid 3-D objects_ can have a _plane of symmetry_.

A plane mirror surface can be drawn through, _but the shape must be exactly the same on both sides of the plane_ (i.e. mirror images), like these are:

Planes of Symmetry

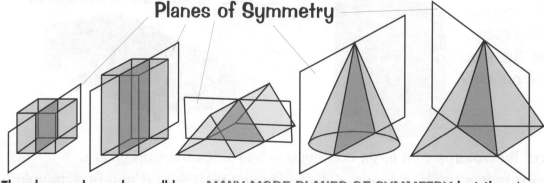

The shapes drawn here all have MANY MORE PLANES OF SYMMETRY but there's only one plane of symmetry drawn in for each shape, because otherwise it would all get really messy and you wouldn't be able to see anything.

SECTION TWO — SHAPES

Symmetry

3) _Rotational Symmetry_

This is where you can <u>ROTATE</u> the shape or drawing into different positions that <u>all look exactly the same</u>.

Order 1

Order 2

Order 3

Order 4

The <u>ORDER OF ROTATIONAL SYMMETRY</u> is the posh way of saying:
"<u>HOW MANY DIFFERENT POSITIONS LOOK THE SAME</u>".
E.g. You should say the S shape above has _"Rotational symmetry order 2"_
BUT...when a shape has <u>ONLY 1 POSITION</u> you can _either_ say that it has
"Rotational Symmetry order 1"   _or_ that it has   _"<u>NO Rotational Symmetry</u>"_

Tracing Paper — this always makes symmetry a lot easier

1) For <u>REFLECTIONS</u>, trace one side of the drawing and the mirror line too.
   Then _turn the paper over and line up the mirror line_ in its original position.

2) For <u>ROTATIONS</u>, just swizzle the paper round. It's really good for _finding the centre of rotation_ (by trial and error) as well as the _order of rotational symmetry_.

3) You can use tracing paper in the <u>EXAM</u> — ask for it, or take your own in.

Tessellations — "Tiling patterns with no gaps"

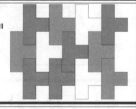

You must have done loads of these, but don't forget
what the name _"tessellation"_ means — _"a tiling pattern with no gaps"_:

The Acid Test

1) Copy these letters and mark in all the <u>lines of symmetry</u>. Also say what
   the <u>rotational symmetry</u> is for each one.

   T   **K**   I   N   S   M

2) Copy all the five solids on the last page <u>without their plane of symmetry</u>
   (see P. 33). Then draw in a <u>different</u> plane of symmetry for each one.

   (Drawing 3-D objects ain't easy but it's good laughing at everyone else's dismal efforts.)

The Shapes You Need to Know

These are easy marks in the Exam — Make sure you know them all!

1) SQUARE

4 lines of symmetry
Rotational symmetry order 4

AS SHOWN ⇒

2) RECTANGLE

2 lines of symmetry
Rotational symmetry order 2

AS SHOWN ⇒

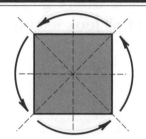

3) RHOMBUS

A square pushed over:

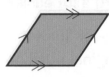 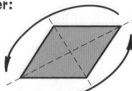

A rhombus is actually a diamond shape but in the Exam you must always call it a rhombus

2 lines of symmetry
Rotational symmetry
order 2
⇐ AS SHOWN

4) PARALLELOGRAM

A rectangle pushed over — two pairs of parallel sides:

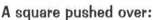

NO lines of symmetry

Rotational symmetry order 2   AS SHOWN ⇑

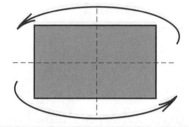

5) TRAPEZIUM

These have One pair of parallel sides

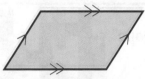

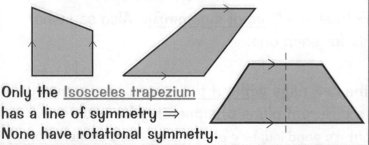

Only the Isosceles trapezium
has a line of symmetry ⇒
None have rotational symmetry.

6) KITE

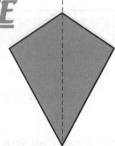

1 line of symmetry
No rotational symmetry

The Shapes You Need to Know

7) EQUILATERAL Triangle

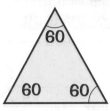

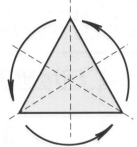

3 lines of symmetry
Rotational symmetry order 3

8) SCALENE Triangle

All three sides different,
All three angles different.

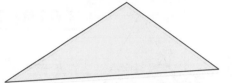

No symmetry, pretty obviously.

9) RIGHT-ANGLED Triangle

No symmetry unless the angles are 45° as shown:

45°   45°

In which case there is one line of symmetry

10) ISOSCELES Triangle

2 sides equal
2 angles equal

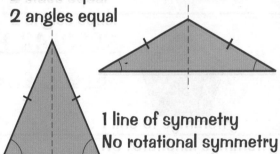

1 line of symmetry
No rotational symmetry

11) SOLIDS

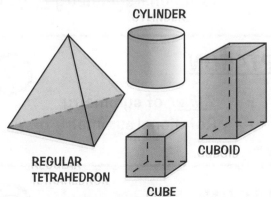

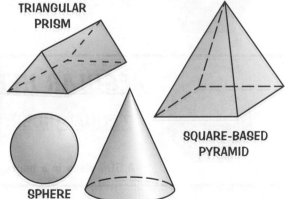

CYLINDER

TRIANGULAR PRISM

REGULAR TETRAHEDRON

CUBE

CUBOID

SPHERE

CONE

SQUARE-BASED PYRAMID

The Acid Test

LEARN everything on these two pages.

Then turn over and write down all the details that you can remember. Then try again.

Regular Polygons

1) A <u>POLYGON</u> is just a fancy word for a *shape with lots of sides*.
2) A <u>REGULAR</u> POLYGON is one where *all the sides and angles are the SAME*.
3) The <u>REGULAR POLYGONS</u> are a *never-ending* series of shapes with some fancy features and *they're very easy to learn*.
4) *Here are the first few but they don't stop* — you can have one with 10 sides or 16 sides or 25 sides etc. <u>THE SYMMETRY IS ALWAYS OBVIOUS.</u>

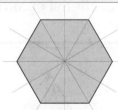

EQUILATERAL TRIANGLE   ③

<u>3 EQUAL SIDES</u>:   *3 lines* of symmetry
Rotational symmetry *order 3*

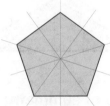

SQUARE   ④

<u>4 EQUAL SIDES</u>:   *4 lines* of symmetry
Rotational symmetry *order 4*

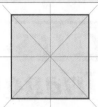

REGULAR PENTAGON   ⑤

<u>5 EQUAL SIDES</u>:   *5 lines* of symmetry
Rotational symmetry *order 5*

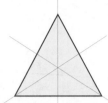

REGULAR HEXAGON   ⑥

<u>6 EQUAL SIDES</u>:   *6 lines* of symmetry
Rotational symmetry *order 6*

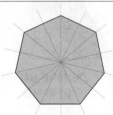

REGULAR HEPTAGON   ⑦

<u>7 EQUAL SIDES</u>:   *7 lines* of symmetry
Rotational symmetry *order 7*

A 50p piece is a heptagon.

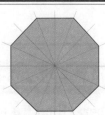

REGULAR OCTAGON   ⑧

<u>8 EQUAL SIDES</u>:   *8 lines* of symmetry
Rotational symmetry *order 8*

Regular Polygons

Interior And Exterior Angles

If you get a regular polygon in the Exam, you're bound to have to work out the interior and exterior angles, so make sure you learn how to do it!

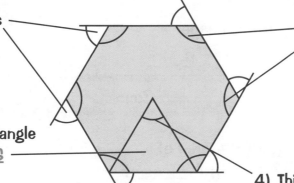

1) Exterior Angles

2) Interior Angles

Did someone say angel?

3) Each sector triangle is ISOSCELES

4) This angle is always the same as the Exterior Angles

LEARN THESE TWO FORMULAS:

$$\text{EXTERIOR ANGLE} = \frac{360°}{\text{No. of Sides}}$$

$$\text{INTERIOR ANGLE} = 180° - \text{EXTERIOR ANGLE}$$

REMEMBER you ALWAYS have to find the *Exterior angle* first, before you can find the Interior angle.

Example

"Calculate angles A and B in the diagram below."

ANSWER: Angle A is an Exterior angle, so we use the formula:

$$\text{Exterior Angle} = \frac{360°}{n} = \frac{360°}{6} = 60° \quad \text{so } \underline{A = 60°}$$

Angle B is an INTERIOR ANGLE = 180° − Exterior Angle
= 180° − 60° = 120°   (= Angle B)

The Acid Test

LEARN THESE TWO PAGES. Then answer these:

1) What is a Regular Polygon?
2) Draw a Pentagon and an Hexagon and put in all their lines of symmetry.
3) Write down the formulas for Exterior and Interior angles.
4) Work out the two angles A and B for the shape shown here:

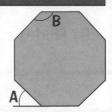

Revision Test for Section Two

WHAT YOU'RE SUPPOSED TO DO HERE is put all the methods
of Section Two into practice to answer these questions.

Revision Test

1) What is a *perimeter*?
   Find the *perimeter* of this shape:

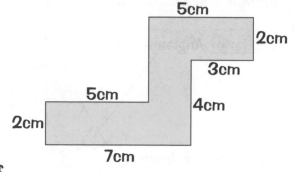

2) What is the *FORMULA* for the *area* of
   a) A *rectangle*?   b) A *triangle*?

3) What are the *TWO CIRCLE FORMULAS*?

4) *Draw a circle* and show on it the *RADIUS*, the *DIAMETER* and the
   CIRCUMFERENCE.

5) Do the *FULL PROPER METHOD* to work out the *area* of these shapes:

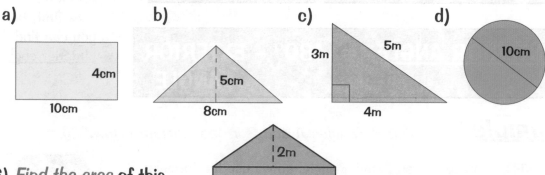

a)   b)   c)   d)

6) *Find the area* of this
   shape:

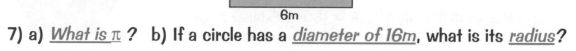

7) a) *What is* π ?   b) If a circle has a *diameter of 16m*, what is its *radius*?

8) A cherry pie has a *radius of 12cm*. Find its *CIRCUMFERENCE*.

9) Find the *AREA* of the same pie.

Revision Test for Section Two

10) A coach has a wheel of _radius 0.6m_. How many times will
it need to rotate for the coach to _move forward 100m_?

11) Do a _sketch_ of these four solids and _draw the net_ for each one:
a) A _cube_ b) A _cuboid_ c) A _triangular prism_ d) A _pyramid_

12) _Work out the volume_ of these two objects:

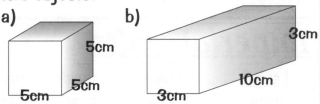

a) b)

13) What are the _three types of symmetry_?

14) _Draw these 7 shapes_ with _all_ their _lines of symmetry_:

a) _Parallelogram_ b) _Rhombus_ c) _Trapezium_
d) _Kite_ e) _Isosceles Triangle_
f) _Equilateral Triangle_ g) _Right-angled Triangle_

Also say what _rotational symmetry_ they have.

15) Sketch these four shapes and show _one plane of symmetry_ on each:

a) _Cuboid_ b) _Triangular Prism_ c) _Cone_ d) _Cylinder_

16) What is a _Regular Polygon_? Draw _two different ones_ and name them.

17) Shown here is a _Regular Pentagon_ (5 sides).
Work out the _EXTERIOR_ and _INTERIOR angles_ for it.

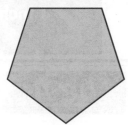

18) What is a _tessellation_?
Draw one on squared paper
using this shape:

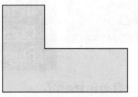

SO IF YOU GET STUCK with any of these questions, _have a look
back at the right page_ in Section Two to _find out how to do it_!

Metric and Imperial Units

This topic is Easy Marks! — make sure you get them.

Metric Units

1) Length   mm, cm, m, km
2) Area mm^2, cm^2, m^2, km^2,
3) Volume mm^3, cm^3, m^3,
              litres, ml
4) Weight   g, kg, tonnes
5) Speed   km/h, m/s

MEMORISE THESE KEY FACTS:

1cm = 10mm	1 tonne = 1000kg
1m = 100cm	1 litre = 1000ml
1km = 1000m	1 litre = $1000cm^3$
1kg = 1000g	$1 cm^3$ = 1 ml

Imperial Units

1) Length   Inches, feet, yards, miles
2) Area   Square inches, square feet, square yards, square miles
3) Volume   Cubic inches, cubic feet, gallons, pints
4) Weight   ounces, pounds, stones, tons
5) Speed   mph

LEARN THESE TOO!

1 Foot = 12 Inches
1 Yard = 3 Feet
1 Gallon = 8 Pints
1 Stone = 14 Pounds (lbs)
1 Pound = 16 Ounces (Oz)

Metric-Imperial Conversions

YOU NEED TO LEARN THESE – they DON'T promise to give you these in the Exam and if they're feeling mean (as they often are), they won't.

Approximate Conversions

1 kg = 2 ¼ lbs   1 gallon = 4.5 litres
1m = 1 yard (+ 10%)   1 foot = 30cm
1 litre = 1 ¾ Pints   1 metric tonne = 1 imperial ton
1 Inch = 2.5 cm   1 mile = 1.6km
                           or 5 miles = 8 km

The Acid Test

In the shaded boxes above, there are 21 Conversions. LEARN THEM, then turn the page over and write them down.

1) a) How many cm is 3 metres?   b) How many mm is 4cm?
2 a) How many kg is 1500g?   b) How many litres is 2000 cm^3?
3) A rod is 40 inches long. What is this in feet and inches?
4) a) Roughly how many yards is 100m? b) How many cm is 6 feet?

Rounding off Measurements

A lot of things you measure have a value *which you can never know exactly*, *no matter how carefully you try and measure them*.
Take this tadpole for example:

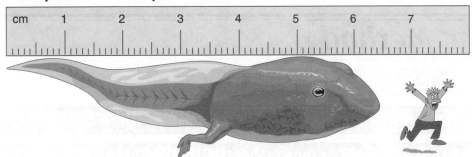

It has a length somewhere *between 6cm and 7cm* and if you look closer you can even say it's somewhere *between 6.4cm and 6.5cm, but you can't really tell any more accurately than that*.

So really *we only know its length to within 0.1cm*. (But let's face it, who needs to know the length of a tadpole more accurately than that?)

The thing is though that *whenever you measure such things as lengths, weights, speeds etc*, you always have to take your answer to *a certain level of accuracy* because *you can never get the exact answer*.

 The simple rule is:

> ### You always round off to the number that it's NEAREST TO

If we take our tadpole, then *to the nearest cm* his length is *6cm* (rather than 7cm) and *to the nearest 0.1cm* it's *6.4cm* (rather than 6.5)

Postage Rates

All this about rounding off to the nearest number is all well and good, but in the Exam they can cheerfully spring a question on you about how much it costs to send a parcel and then the rules are completely different.

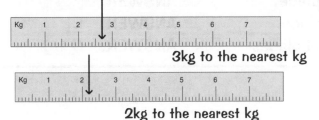

3kg to the nearest kg

2kg to the nearest kg

POSTAGE RATES	
Weight not over:	Price
1kg	£2.70
2kg	£3.30
3kg	£4.10

One reading would round off to 3kg and the other to 2kg but the postage rate is set by "WEIGHT NOT OVER.." so *both of these would cost the same* (£4.10) because they are both more than 2kg but less than 3kg. It's tricky, so watch out for questions like that.

SECTION THREE — MORE NUMBERS

Rounding Off

When you have <u>DECIMAL NUMBERS</u> you might have to round them off to the nearest *whole number*. The trouble is, they could also ask you to round them off to either <u>ONE DECIMAL PLACE</u> (or possibly <u>TWO</u> *decimal places*). This isn't too bad but you do have to learn some rules for it:

Basic *Method*

1) <u>Identify</u> the position of the LAST DIGIT.

2) Then look at the <u>next digit to the right</u> – called the DECIDER.

3) If the DECIDER is <u>5 or more</u>, then <u>ROUND-UP</u> the LAST DIGIT.
If the DECIDER is <u>4 or less</u>, then leave the LAST DIGIT as it is.

<u>EXAMPLE:</u>   What is 3.65 to 1 Decimal Place?

$$3.65 \quad = \quad 3.7$$

<u>LAST DIGIT</u> to be written
(because we're rounding to 1
Decimal Place)

DECIDER

The *LAST DIGIT*
ROUNDS UP to 7
because the *DECIDER*
is *5 or more*

Decimal Places (D. P.)

1) To round off to <u>ONE DECIMAL PLACE</u>, the *LAST DIGIT* will be the
   one *just after the decimal point*.
2) There must be <u>NO MORE DIGITS</u> after the *LAST DIGIT* (not even zeros).

<u>EXAMPLES</u>
Round off 5.14 to 1 decimal place.   ANSWER: <u>5.1</u>
Round off 2.37 to 1 decimal place.   ANSWER: <u>2.4</u>
Round off 1.08 to 1 decimal place.   ANSWER: <u>1.1</u>
Round off 3.846 to 2 decimal places   ANSWER: <u>3.85</u>

The Acid Test

1) LEARN the <u>3 Steps of the Basic Method</u> and the <u>2 Extra Rules</u> for Decimal Places.
2) Round these numbers off to <u>1 decimal place</u>:
     a) 2.34   b) 4.56   c) 3.31   d) 9.85   e) 0.76
3) Round these off to the <u>nearest whole number</u>:
     a) 1.91   b) 2.1   c) 4.5   d) 0.9   e) 5.28

Rounding Off

Rounding Whole Numbers

The easiest ways to round off a number are:

1) "To the nearest WHOLE NUMBER"   3) "To the nearest HUNDRED".
2) "To the nearest TEN"   4) "To the nearest THOUSAND"

This isn't difficult so long as you remember the *2 RULES*:

> 1) The number always lies between 2 POSSIBLE ANSWERS,
>   Just choose the one it's NEAREST TO.
>
> 2) If the number is exactly in the MIDDLE,
>   then ROUND IT UP.

EXAMPLES:

1) Give 581 to the nearest TEN.
   ANSWER: 581 is between 580 and 590, but it is nearer to 580

2) Give 235 to the nearest HUNDRED.
   ANSWER: 235 is between 200 and 300, but it is nearer to 200

3) Round 78.7 to the nearest WHOLE NUMBER.
   ANSWER: 78.7 is between 78 and 79, but it is nearer to 79

4) Round 1500 to the nearest THOUSAND.
   ANSWER: 1500 is between 1000 and 2000. In fact it is *exactly halfway* between them. *So we ROUND IT UP* (see Rule 2 above) to 2000

Significant Figures

> 1) The MORE SIGNIFICANT FIGURES a number has, the MORE ACCURATE it is.
> 2) The NUMBER OF SIGNIFICANT FIGURES is just HOW MANY DIGITS the number has at the front THAT ARE NOT ZERO.

EXAMPLES:
   117 has 3 significant figures   120 has 2 sig fig   300 has 1 sig fig
   9950 has 3 sig fig.   7000 has 1 sig fig   4.6 has 2 sig fig

The Acid Test

1) Round these off to the nearest 10:
   a) 776   b) 594   c) 44.1   d) 27   e) 98

2) Round these numbers off to the nearest hundred: a) 3626   b) 750   c) 256

Estimating and Approximating

1) Estimating

This is quite easy. To _estimate_ something this is all you do:

> 1) ROUND EVERYTHING OFF to nice easy CONVENIENT NUMBERS
> 2) Then WORK OUT THE ANSWER using these nice easy numbers

1) Don't worry about the answer being "wrong".
2) You're only trying to get a _rough idea_ of the size of the proper answer.
   In other words, is it about 20 or about 200, for example?
3) In the Exam you'll need to _show all the steps you've done_, to prove you
   didn't just use a calculator.

> EXAMPLE   _"Estimate the value of $\frac{68 + 21}{45.3}$ and show all your working out"_
>
> Ans: $\frac{68 + 21}{45.3} \approx \frac{70 + 20}{45} = \frac{90}{45} = 2$ ("$\approx$" means _roughly equal to_)

2) Estimating Areas

This is easier than you might think too. All you have to do is:

> Round off all lengths to the NEAREST WHOLE, and work it out — easy.

Estimate the <u>area of this rectangle</u>:

The _area of the rectangle_
is _roughly_ 2m by 3m
        = roughly <u>6m²</u>.

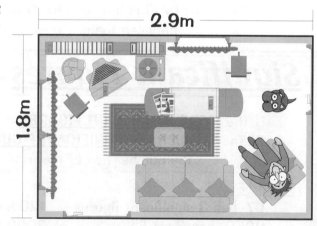

2.9m

1.8m

The Acid Test

1) Without using your calculator, estimate the answer to this: $\frac{131+52}{59.3}$
2) Estimate the area of the palm of your hand in cm².
3) What is the approximate area of your classroom floor, in m²?

Conversion Graphs

THESE ARE REALLY EASY. I know graphs are generally a bit grim but *Conversions Graphs* are OK. *In the Exam* you're likely to get a Conversion Graph question which converts between things like these:

£ → Dollars,	£ → Francs,	£ → DMarks, etc
Pints → Litres,	Gallons → Litres,	etc
Miles → Kilometres,	mph → km/h,	etc

In fact, it hardly matters what the conversion is, because the method is exactly the same every time — and very easy to remember.

An Important *EXAMPLE*:

This graph converts between miles and kilometres

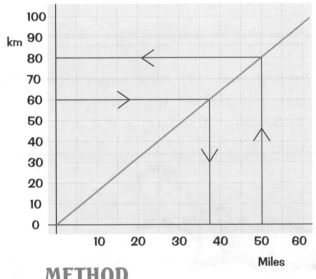

Miles

2 VERY TYPICAL QUESTIONS:

1) How many miles is 60 km?

ANS: Draw a line *straight across* from "60" on the "km" axis 'till it *hits the line*, then go *straight down* to the "miles" axis and read off the answer: <u>37.5 miles</u>.

2) How many km is 50 miles?

ANS: Draw a line *straight up* from "50" on the "miles" axis 'till it *hits the line*, then go *straight across* to the "km" axis and read off the answer: <u>80 km</u>.

METHOD

> 1) <u>Draw a line</u> from the <u>value</u> on one axis.
>
> 2) Keep going 'till you <u>hit the LINE</u>.
>
> 3) Then <u>change direction</u> and go straight to <u>the other axis</u>.
>
> 4) <u>Read off the new value</u> from the axis. <u>That's the answer</u>.

If you remember those 4 simple steps you really can't go wrong — let's face it, conversion graphs are a doddle.

The Acid Test | LEARN the 4 steps for using Conversion Graphs

From the above Conversion Graph,
1) Find how many km is equivalent to   a) 25 miles;   b) 45 miles.
2) Find how many miles are equal to   a) 20km;   b) 50km.

Conversion Factors

Conversion Factors are a really good way of dealing with all sorts of questions and the method is dead easy.

Method

1) Find the <u>Conversion Factor</u>   (always easy)

2) <u>Multiply AND divide by it</u>

3) Choose the <u>common sense answer</u>

Example 1

"The Giant pond gremlin is a rare occurrence these days. This one, called Basil, was discovered recently. He was over 5.85m in length.   How long is this in cm?"

"He was over 5.85m in length.."

<u>Convert 5.85m into cm</u>.

Step 1) <u>Find the CONVERSION FACTOR</u>

In this question the Conversion factor = <u>100</u>
— simply because 1m = <u>100</u> cm

Step 2) <u>MULTIPLY AND DIVIDE by the conversion factor</u>:

5.85m × 100 = 585 cm (makes sense)

5.85 m ÷ 100 = 0.0585 cm (ridiculous)

Step 3) <u>Choose the COMMON SENSE answer</u>:

Obviously the answer is that 5.85m = <u>585 cm</u>

Conversion Factors

Example 2

"If £1 is equal to 2.75 Deutschmarks, how much is 20 DM in £ and p?"

Step 1) Find the CONVERSION FACTOR

In this question the *Conversion Factor is* obviously 2.75

(When you're changing foreign money it's called the "Exchange Rate")

Step 2) MULTIPLY AND DIVIDE by the conversion factor:

$$20 \times 2.75 = 55 = £55$$
$$20 \div 2.75 = 7.27 = £7.27$$

Step 3) Choose the COMMON SENSE answer:

Not quite so obvious this time, but if roughly 3 DM = £1,

then 20 DM can't be much — certainly not £55,

so the answer must be   £7.27p

Example 3

"A popular item at our local Supplies is "Froggatt's Hot Sickly Sauce" (not available in all areas). The Farmhouse Economy Size is the most popular and weighs 1500g. How much is this in kg?"

Step 1)   Conversion Factor = 1000   (simply because 1kg = 1000g)

Step 2) $1500 \times 1000 = 1,500,000 \, kg$ (Uulp..)

$1500 \div 1000 = 1.5 \, kg$ (That's more like it)

Step 3)   So the answer must be that 1500 g = 1.5 kg

The Acid Test

LEARN the 3 steps of the Conversion Factor method. Then turn over and write them down.

1) Basil the pond gremlin was found to weigh 0.12 Tonnes. What's this in kg?

2) Froggatt's also do a nice line in canned weak tea. How many pints is 1½ gallons?

Fractions

The Fraction Button: $a\frac{b}{c}$ Use this as much as possible

It's very easy, so make sure you know how to use it:

1) Entering Fractions into the Calculator

E.g.   To enter ⅓ press $\boxed{1}$ $\boxed{a\frac{b}{c}}$ $\boxed{3}$

To enter 1⅖ press $\boxed{1}$ $\boxed{a\frac{b}{c}}$ $\boxed{2}$ $\boxed{a\frac{b}{c}}$ $\boxed{5}$

2) Converting Fractions to Decimals

Enter the fraction, then press $\boxed{=}$, then press $\boxed{a\frac{b}{c}}$ repeatedly

E.g. press $\boxed{1}$ $\boxed{a\frac{b}{c}}$ $\boxed{4}$ $\boxed{=}$ $\boxed{a\frac{b}{c}}$ $\boxed{a\frac{b}{c}}$ $\boxed{a\frac{b}{c}}$.... The display keeps swapping

between ¼ and 0.25,   i.e. between fraction and decimal.

3) Calculations With Fractions: "Of" means "×"

> Remember: "of" means "×"

E.g. To work out ⅕ of £90,   you'd say to yourself: ⅕ × £90

So you'd press $\boxed{1}$ $\boxed{a\frac{b}{c}}$ $\boxed{5}$ $\boxed{×}$ $\boxed{90}$ $\boxed{=}$ which gives £18.

4) To Reduce a Fraction to its Lowest Terms

Simply ENTER IT and then press $\boxed{=}$

E.g. to cancel ¹²⁄₁₆ down to its lowest terms:

$\boxed{12}$ $\boxed{a\frac{b}{c}}$ $\boxed{16}$ $\boxed{=}$ $\boxed{3 \lrcorner 4}$ = ¾

5) Patterns in Fractions

> Find the MULTIPLIER to get from one fraction to the next, then apply it to the other number

EXAMPLE:

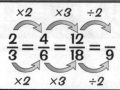

$$\underset{\times2}{\overset{\times2}{\frac{2}{3}}}=\underset{\times3}{\overset{\times3}{\frac{4}{6}}}=\underset{\div2}{\overset{\div2}{\frac{12}{18}}}=\frac{}{9}$$

Between the first two fractions the multiplier is 2. From the 2nd to the 3rd fraction is "×3" so the top number must become 12. From the 3rd to the 4th it's ÷2 so the top number must become 6.

The Acid Test

1) Use your Fraction Button to give ⅗ as a decimal.

2) Use your Fraction Button to work out ⅚ of £720.

3) Use your Fraction Button to cancel these fractions to their simplest

terms: a) ⁶⁄₉;   b) ⁶⁰⁄₇₂;   c) ³⁵⁄₄₉.

Fractions, Decimals and Percentages

1) *Fractions, Decimals and Percentages* are just

different ways of saying the same thing — ½, 0.5 and 50% *all mean the same thing, don't they*. These three *you should know straight off* without any problem:

Diagram	Fraction	Decimal	Percentage
¼	¼	0.25	25%
½	½	0.50	50%
¾	¾	0.75	75%

2) *Converting Fractions to Decimals to Percentages*

For the ones you don't know, you must be able to convert them, like this:

Fraction → Divide using the calculator → **Decimal** → × by 100 → **Percentage**

E.g. $\frac{1}{5}$ (1 ÷ 5) = **0.2** (0.2×100) = **20%**

EXAMPLE: Convert $\frac{1}{8}$ to a decimal and to a percentage.

Answer: Fraction → Decimal → Percentage

$\frac{1}{8}$ (1 ÷ 8) = **0.125** (0.125 × 100) = **12.5%**

The Acid Test

Learn all the 3 rows in the table above and then fill in the table below:

Fraction	Decimal	Percentage
1/5		
		40%
	0.8	
1/10		
	0.7	
		37½%
5/8		

Percentages

Discounts, VAT, Interest, Increase, Etc

<u>MOST</u> percentage questions are like this:

> Work out "something %" of "something else"

E.g. Find **20%** of **£80**

This is the method to use:

1) <u>WRITE IT DOWN</u>:   Find   **20%**   **of**   **£80**

2) <u>TRANSLATE IT INTO MATHS</u>: $\dfrac{20}{100}$ $\times$ 80

3) <u>WORK IT OUT</u>:   20 ÷ 100 × 80 = **£16**

TWO IMPORTANT DETAILS:

Make sure you remember them!

1) <u>"Per cent" means "out of 100"</u>

so <u>20%</u> means "20 out of 100" = <u>20 ÷ 100</u> = $\dfrac{20}{100}$

(That's how you work it out in the method shown above)

2) <u>"OF" means "×"</u>

In maths, the word "of" can always be replaced with "×" for working out the answer
(as shown in the above method)

Percentages

Important **Example** No. 1

1) A calculator is priced at £15 but there is a discount of 30% available.
FIND THE REDUCED PRICE OF THE CALCULATOR.

Answer: First find 30% of £15 using the method from the last page:

1)   30%   of   £15

2) $\frac{30}{100}$ ×   15

3) $\boxed{30} \div \boxed{100} \times \boxed{15} = $ 4.5   = £4.50

See P.17 if you don't know why 4.5 is £4.50

This is the <u>DISCOUNT</u> so we _take it away_ to get the final answer:

$$£15 - £4.50 = \underline{£10.50}$$

Important **Example** No. 2

2)   A joiner's bill for fitting a kitchen is £1800 + VAT.
The VAT is charged at 17.5%.
WORK OUT THE TOTAL BILL.

Answer: First find 17.5% of £1800 using the standard method:

1)   17.5%   of   £1800

2) $\frac{17.5}{100}$ ×   1800

3) $\boxed{17.5} \div \boxed{100} \times \boxed{1800} = $ 315 = £315

This £315 is the VAT which then _has to be ADDED_ to the £1800 to give the _FINAL BILL_:

$$£1800 + £315 = \underline{£2115}$$

Percentages

Comparing Numbers using Percentages

This is the other common type of percentage question.

> # Give "one number"
>
> ## AS A PERCENTAGE OF
>
> # "another number"

For example, "Express £5 <u>as a percentage of</u> £50."

This is the method to use:

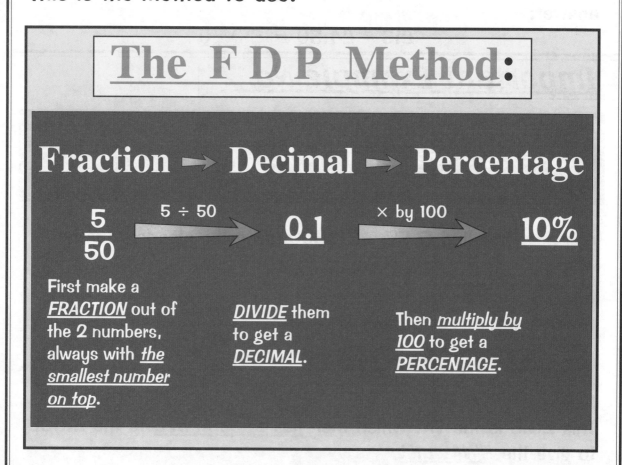

The F D P Method:

Fraction → Decimal → Percentage

$$\frac{5}{50} \xrightarrow{5 \div 50} \underline{0.1} \xrightarrow{\times \text{ by } 100} \underline{10\%}$$

First make a *FRACTION* out of the 2 numbers, always with *the smallest number on top.*

DIVIDE them to get a *DECIMAL*.

Then *multiply by 100* to get a *PERCENTAGE*.

Percentages

Two Important Examples

1) *"A shopkeeper buys watches at £4 each and sells them for £5 each. What is his profit AS A PERCENTAGE?"*

Answer: The two numbers we want to *compare* are the PROFIT (which is £1) with the ORIGINAL cost (which is £4). We then apply the FDP method:

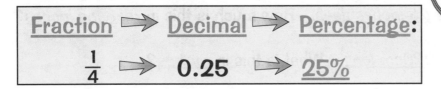

Fraction $\Rightarrow$ Decimal $\Rightarrow$ Percentage:

$\frac{1}{4}$ $\Rightarrow$ 0.25 $\Rightarrow$ 25%

so the shopkeeper makes a 25% profit on the watches.

2) *"In a sale, a computer game is reduced in price from £80 to £64. What PERCENTAGE REDUCTION is this?"*

Answer: The two numbers we want to *compare* are the REDUCTION (which is £16) and the ORIGINAL VALUE (which is £80). We then apply the FDP method:

Fraction $\Rightarrow$ Decimal $\Rightarrow$ Percentage:

$\frac{16}{80}$ $\Rightarrow$ 0.2 $\Rightarrow$ 20%

The Acid Test

For these two questions, decide which type they are and use the right method.

1) A bank charges interest at 7% per year. If £1000 is borrowed for one year, how much interest will be charged?

2) A house decreases in value from £55,000 to £50,000. What is the decrease in the value of the house as a percentage?

Revision Test for Section Three

WHAT YOU'RE SUPPOSED TO DO HERE is put all the methods of Section Three into practice to answer these questions.

Revision Test

1) a) How many _cm_ are there in a _metre?_
    b) How many _metres_ are there in a _km?_
    c) How many _ml_ are there in a _litre?_
    d) How many _g_ are there in a _kg?_

2) A chair is _26 inches deep_.   How much is this in _feet and inches?_

3) A pipe is _220cm long_.   What is this in _metres?_

4) How many _cm³_ are there in _1.6 litres?_

5) Round these numbers off _to 1 decimal place_:

    a) 6.41   b) 5.46   c) 8.25

6) These are the sort of numbers you might get in your _calculator display_:
    a) 1.7272727   b) 2.3888888   c) 5.35483871   d) 10.875

                    Round them off to the _nearest whole number_.

7) a) Give 526 to the _nearest 10_   b) Give 493 to the _nearest 100_

8) a) Give 3360 to the _nearest 1000_   b) Give 45 to the _nearest 10_

9) How many _significant figures_ have these numbers got:
    a) 19   b) 350   c) 1100   d) 45.2   e) 6,000

10) _Without using your calculator_,

                    ESTIMATE the answer to $\dfrac{510}{18.1 + 32}$

11) _Estimate_ the height of a two-storey house, _in metres_.

Revision Test for Section Three

12) *Estimate* the *area* of a square rug with side length 1.9m.

13) What are the *Three Steps* of the *Conversion Factor Method*?

14) Use them to *convert 3.75 m into cm*.

15) If the *Exchange Rate* is *£1 = 8.5 French Francs*, use the conversion factor method to find *how many £* it is for *68 Francs*. *Also find* how many Francs it is for *£20*.

16) *1 mile = 1.6 km*. How many miles is *34 km*? Give it to the *nearest mile*.

17) *1 Tonne = 1000kg*. How many kg is *2.59 Tonnes*?

18) What is the ⓐᵇ_c button for? Use it to *work out* a) ⅕ of £85
  b) ⅓ of £120

19) Use the ⓐᵇ_c button to a) *convert* ⅛ into a *decimal*
  b) *reduce* ¹⁸⁄₂₄ to its *simplest form*.

20) How do you convert from *fraction to decimal*? How do you convert from *decimal to percentage*?

21) Convert ½ into a *decimal* and then into a *percentage*.

22) Convert ⅕ into a *decimal* and then into a *percentage*.

23) A cassette player is priced at £28 but there is a *25% discount*. Find how much you will *save* with the discount and how much the cassette player will then *cost*.

24) In another shop a coat is *reduced* from *£95 to £76*. What *percentage reduction* is this?

SO IF YOU GET STUCK with any of these questions, have a look back at the right page in Section Three to find out how to do it!

Mean, Median, Mode and Range

If you don't manage to _learn these 4 basic definitions_ then you'll be passing up on some of the easiest marks in the whole Exam.   _It can't be that difficult can it?_

1) _MODE_   =   _MOST_ common

2) _MEDIAN_   =   _MIDDLE_ value

I'm mean.

3) _MEAN_   =   _TOTAL of items_ _÷ NUMBER of items_

4) _RANGE_   =   How far from the _smallest_ to the _biggest_

The Golden Rule

Mean, median and mode should be _easy marks_ but even people who've gone to the incredible extent of learning them still manage to lose marks in the Exam because they don't do _this one vital step:_

Always REARRANGE the data in ORDER OF SIZE

(and check you have the same number of entries as before!)

Mean, Median, Mode and Range

Example Find the mean, median, mode and range of these numbers:

4, 9, 2, 3, 2, 2, 5, 1, 7   (9 numbers)

1) FIRST... rearrange them: 1, 2, 2, 2, 3, 4, 5, 7, 9   (✓ still 9)

2) MEAN = $\frac{\text{total}}{\text{number}}$ = $\frac{1+2+2+2+3+4+5+7+9}{9}$

= 35 ÷ 9 = <u>3.89</u>

3) MEDIAN = <u>the middle value</u>
(only when they are <u>arranged in order of size</u>, that is!)

(<u>When there are two middle numbers</u> the median is **HALFWAY BETWEEN THE TWO MIDDLE NUMBERS**)

1, 2, 2, 2, 3, 4, 5, 7, 9
← four numbers this side ↑ four numbers this side →
Median = <u>3</u>

4) MODE = <u>most</u> common value, which is simply <u>2</u>.

5) RANGE = distance from lowest to highest value,
i.e. from 1 up to 9, = <u>8</u>

Remember:

<u>Mo</u>de = <u>mo</u>st (emphasise the 'o' in each when you say them)
<u>M*d</u>ian = <u>m*d</u> (emphasise the m*d in each when you say them)
<u>Mean</u> is just the <u>average</u>, but it's <u>mean</u> 'cos you have to work it out!

The Acid Test

LEARN The Four Definitions and *The Golden Rule...*

..then turn this page over and <u>write them down from memory</u>.
Then find the mean, median, mode and range for this set of data:
5, 10, 16, 3, 8, 11, 7, 10, 2

Tally/Frequency Tables

Why bother doing a tally?
Answer: — *TO GET IT RIGHT, basically!*

If you try to fill in the frequency table <u>WITHOUT DOING THE TALLY</u> you're bound to mess it up — and that means <u>YOU'LL LOSE LOADS OF EASY MARKS</u>. It's as simple as that. By the way, the word <u>FREQUENCY</u> just means "<u>how many</u>" so a <u>frequency table</u> is just a "<u>HOW MANY IN EACH GROUP</u>" table.

Example:
The registration letter of the cars in a school car park were:

N̶ S̶ R̶ L̶ N̶ M R J N
P M L N P R R N

The tally has been done for the first 5 letters in the above list:

Registration letter	Tally	Frequency
J		
K		
L	|	
M		
N	||	
P		
R	|	
S	|	
		Total

Four Important Points

1) DO EACH LETTER OR NUMBER ONE AT A TIME Make a <u>TALLY MARK</u> (in the right box!) for each one and *strike it out* (to show it's been done). *This is the only way you'll get it right*.

2) FILL IN THE LAST COLUMN by adding up each tally —

<u>BUT ONLY WHEN YOU'VE DONE ALL THE LETTERS OR NUMBERS</u> and finished the tally.

3) Now the check: Add up the last column — <u>THE TOTAL</u> should be how many numbers or letters there were to start with. *If not you'll have to do the whole thing again!*

4) In a tally, <u>every 5th mark crosses a group of 4 like this:</u> ⊦⊦⊦⊦

so ⊦⊦⊦⊦ \|\| would represent 7 (a group of 5, plus 2 more).

The Acid Test
Complete the tally/frequency table <u>above</u> and make sure your total is 17. If not <u>do the whole thing again</u>.

Frequency Tables and Bar Charts

Once they've got you to do a frequency table the next thing is to DRAW A BAR CHART.

What Does "$40 \leq w < 60$" Mean?

It basically means _the value of w is between 40 and 60_. But.... you also need to know
1) the $\leq$ symbol means w can be EQUAL TO 40   (or greater than 40)
2) the $<$ symbol means w must be LESS THAN 60   (to go in this group)
The upshot of all this is that _a value of 40_ will go in this group: $40 \leq w < 60$, whereas _a value of 60_ will have to go in the next group up: $60 \leq w < 80$.

An Important Example

25 people in a village were asked their age. The tally table and bar chart here show the result. You'll notice _each number was crossed off_ as it was done:

13, 51, 3, 5, 17, 25, 32, 43, 67, 24, 33, 29, 20,
55, 25, 35, 12, 18, 48, 23, 27, 59, 65, 82, 90

Age A (years)	Tally	Frequency
$0 \leq A < 20$	⊥⊥⊥⊥ |	6
$20 \leq A < 40$	⊥⊥⊥⊥ ⊥⊥⊥⊥	10
$40 \leq A < 60$	⊥⊥⊥⊥	5
$60 \leq A < 80$	| |	2
$80 \leq A < 100$	| |	2

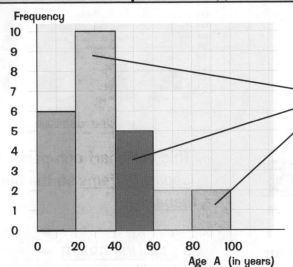

These numbers here are ALWAYS just THE HEIGHTS OF THE BARS in the bar chart.

E.g. there are 5 people between the ages of 40 and 60, so the bar drawn between 40 and 60 goes up to 5.

The Acid Test

Draw a bar chart (like the one above) using the following age data: 24, 2, 51, 76, 9, 83, 44, 18, 16, 22, 29, 30, 55, 33, 40, 80.

Graphs and Charts

1) Pictograms — these use <u>pictures</u> instead of <u>numbers</u>.

<u>EXAMPLE</u>: The *pictogram* opposite shows the number of ice-creams sold at a *3*-screen cinema during Christmas week:

 = 1000 ice-creams

Screen 1	🍦🍦	(2000 delicious ice-creams)
Screen 2	🍦🍦	(1500 delicious ice-creams)
Screen 3	🍦🍦🍦🍦	(3500 delicious ice-creams)

In a PICTOGRAM each picture or symbol represents a certain number of items.

2) Line Graphs or *Frequency Polygons*

A *line graph*, which is sometimes called a *Frequency Polygon*, is just a set of points *joined up with straight lines* like this one →

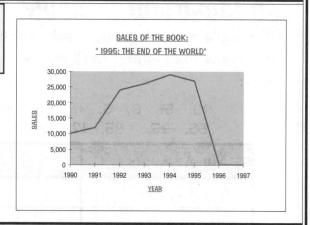

3) Bar Charts Just watch out for when the bars should touch or not touch:

Number of hand spans measured (various lengths)

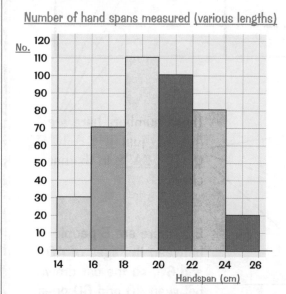

Popular Choices at the School Canteen

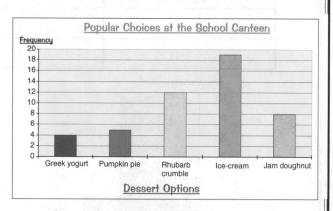

This bar chart compares *totally separate items* so the bars are *separate*.

ALL the bars in this chart are for **LENGTHS** and you must *put every possible length into one bar or the next* so there mustn't be any spaces.

A <u>BAR-LINE GRAPH</u> is just like a bar chart except you just draw thin lines instead of bars.

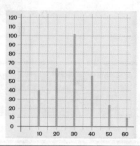

Graphs and Charts

4) Scatter Graphs

1) <u>A SCATTER GRAPH</u> is just a load of points on a graph that *end up in a bit of a mess* rather than in a nice line or curve.

2) There is a posh word to say *how much of a mess* they are in — it's <u>CORRELATION</u>.

3) <u>Good Correlation</u> (or *Strong* Correlation) means the points *form quite a nice line*, and it means *the two things are closely related to each other*.

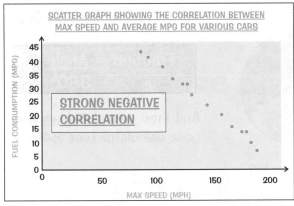

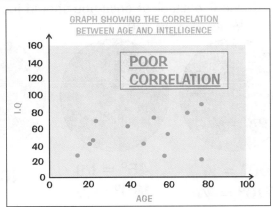

4) <u>Poor Correlation</u> (or *Weak* Correlation) means the points are *all over the place* and so there is *very little relation between the two things*.

5) If the points form a line sloping <u>UPHILL</u> from left to right, then there is <u>POSITIVE CORRELATION</u>, which just means that *both things increase or decrease together*.

6) If the points form a line sloping <u>DOWNHILL</u> from left to right, then there is <u>NEGATIVE CORRELATION</u>, which just means that *as one thing increases the other decreases*.

7) So when you're describing a scatter graph you have to mention both things, i.e. whether it's *strong/weak/moderate* correlation *and* whether it's *positive/negative*.

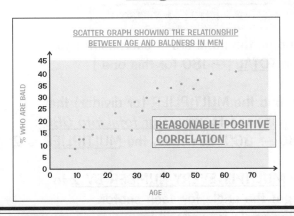

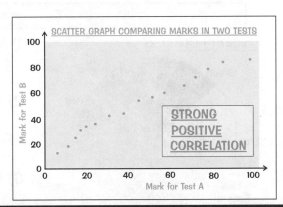

The Acid Test

LEARN ALL THE CHARTS on these two pages

1) Turn over the page and draw an example of each type of chart.

2) If the points on a scatter graph form a line sloping uphill from left to right, what does it tell you about the two things the graph is comparing?

SECTION FOUR — STATISTICS AND GRAPHS

Pie Charts

They can make Pie Charts into quite tricky Exam questions.
So learn the Golden Rule for Pie Charts:

> ## The TOTAL of Everything = 360°

Remember that 360° is the trick for dealing with most Pie Charts

1) Relating Angles to Fractions

These five simplest ones you should just
know straight off:

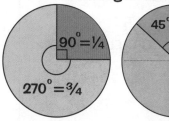 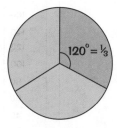

$$90° = ¼$$
$$270° = ¾$$

$$45° = 1/8$$
$$180° = ½$$

$$120° = 1/3$$

For any angle the formula is:

> Fraction = $\dfrac{\text{Angle}}{360°}$

And then *cancel it down* with
your calculator (see P.46)

If you have to measure an angle, you should expect it to be a nice round number like 90°
or 180° or 120°, so don't go writing 89° or 181° or anything silly like that.

2) Relating angles to Numbers of other things

Colour	Green	Red	Yellow	Purple	Blue	Total
Number	24	40	34	30	52	180

×2   ×2

Angle	48°	80°	68°	60°	104°	360°

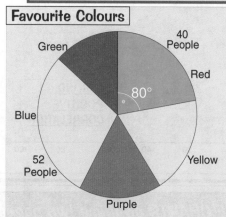

Favourite Colours

1) Add up all the numbers in each sector to
get the TOTAL (← 180 for this one)

2) Then find the MULTIPLIER (or divider) that
you need to *turn your total into 360°*:
For 180 → 360 as above, the MULTIPLIER is 2

3) Now MULTIPLY EVERY NUMBER BY 2 to
get the angle for each sector.
E.g. the angle for red will be
$$40 × 2 = \underline{80°}$$

The Acid Test

Display these data in a Pie Chart:

Rock Stars	Brill	Cool	Sneak	StarZ
No. of Fans	10	30	5	45

Probability

Probability definitely seems a bit of a "Black Art" to most people. It's not as bad you think, but <u>YOU MUST LEARN THE BASIC FACTS</u>, which is what we have on these 3 pages.

I'm getting Saturday's draw...

1) <u>All</u> **Probabilities are** <u>between 0 and 1</u>

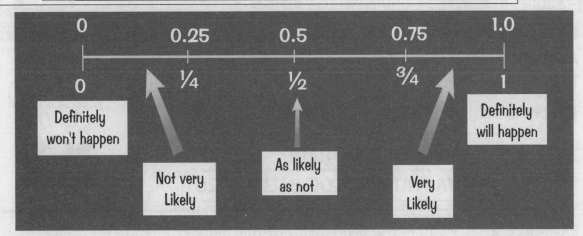

Probabilities can only have values between 0 and 1, and you should be able to put the probability of any event happening on this scale of 0 to 1.

Remember you can give probabilities using either <u>FRACTIONS, DECIMALS</u> or <u>PERCENTAGES</u>.

2) <u>Equal</u> **Probabilities**

When the different results all have the same chance of happening, then the probabilities will be <u>EQUAL</u>. These are the two cases which usually come up in Exams:

1) <u>TOSSING A COIN</u>:   Equal chance of getting a head or a tail (½)

2) <u>THROWING A DICE</u>:   Equal chance of getting any of the numbers (⅙)

Probability

3) *Unequal Probabilities You Can Work Out*

These make for more interesting questions.
    (Which means you'll get them in the Exam.)

> <u>EXAMPLE 1:</u>   *"A bag contains 11 blue balls, 6 red balls and 3 green balls.*
> *Find the probability of picking out a red ball."*

ANSWER:

The chances of picking out the three colours are <u>NOT EQUAL</u>.

The probability of picking a red is simply:

$$\frac{\text{NUMBER OF REDS}}{\text{TOTAL NUMBER OF BALLS}} = \frac{6}{20}$$

<u>EXAMPLE 2</u>:   *"What is the probability of this spinner landing on dots?"*

ANSWER:

The spinner has *the same chance of stopping on every sector*...

... and since there are *2 out of 8 which are dots* then it's a *2 out of 8 chance* of getting dots.

<u>BUT REMEMBER</u> ... you have to say this as a **FRACTION** or a **DECIMAL** or a **PERCENTAGE**:

> 2 out of 8 is 2 ÷ 8 which is <u>0.25</u> (as a decimal)
> or <u>¼</u> (as a fraction) or <u>25%</u> (as a percentage)

4) *Unequal Probabilities You'd Need to Test*

In many *real-life situations* the probabilities are not equal, like in these examples:

1) The probabilities of either *winning, drawing or losing a game* (not 1/3 each!)

2) The chance that *the next car to pass will be red or blue or white etc.*

3) <u>A BIASED DICE</u> coming up with a "Six" compared to any other number.

4) The chance of *passing or failing a test*.

In all these cases *you can only find the probabilities by doing* <u>A TEST OR A</u> <u>SURVEY</u> — they might ask you to say just that in the Exam to see if you know it.

Probability

5) _The Probability of the_ OPPOSITE _Happening is just_ the rest _of the probability_ that's left over

This is simple enough AS LONG AS YOU REMEMBER IT.
If the probability of something happening is say 0.4 then the chance of it NOT HAPPENING is just the rest of the probability that's left over — in this case, 0.6 (so that it adds up to 1).

Example: A dice has a 0.5 chance of coming up with a number less than 4. What is the chance of it _not_ coming up with a number less than 4?

Answer: 1 – 0.5 = 0.5

So, the chance of the dice _not_ coming up with a number less than 4 is 0.5.

6) _Listing_ All Outcomes: 2 Coins, dice, Spinners

A simple question you might get is to list all the possible results from tossing two coins or two spinners or a dice and a spinner, etc. Whatever it is, it'll be very similar to these, so LEARN THEM:

The _possible outcomes_ from TOSSING TWO COINS are:

Head	Head	H H
Head	Tail	H T
Tail	Head	T H
Tail	Tail	T T

From TWO SPINNERS with 3 sides:

BLUE + 1	RED + 1	GREEN + 1
BLUE + 2	RED + 2	GREEN + 2
BLUE + 3	RED + 3	GREEN + 3

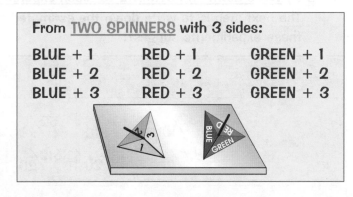

Try and list the possible outcomes METHODICALLY
— to make sure you get them ALL.

The Acid Test

1) If today is a Friday, what is the probability that tomorrow will be a Saturday?

2) A dice is thrown. What is the probability of a) a 4; b) a number other than 4?

3) A box has 4 yellow marbles, 10 red ones and 1 blue one. What is the chance of a yellow marble being picked?

4) List ALL THE POSSIBLE OUTCOMES when a coin and a dice are thrown together.

Probability Experiments

Sometimes you won't be able to calculate a probability, so you'll have to do an experiment instead.

On this page you will see how to estimate a probability from an experiment.

Example: *"A bag contains 20 eyeballs. From experiments, estimate how many blue eyeballs are in the bag".*

Method:

Step 1) Do the EXPERIMENT

Tom, James, Helen and Sarah each do an experiment to try to answer the question. The boys do 100 tries each, and the girls 200. Each "try" involves taking an eyeball out of the bag, noting its colour and putting it back in. The results are:

Tom's: 32 blue eyeballs out of 100 tries.
James':   19 blue eyeballs out of 100 tries.
Helen's: 52 blue eyeballs out of 200 tries.
Sarah's: 46 blue eyeballs out of 200 tries.

Step 2) Write down an ESTIMATE for each experiment

The next step is to write down the estimated number of blue eyeballs which these experiments suggest:

	Tom	James	Helen	Sarah
Result	$\frac{32}{100}$ $=\frac{6.4}{20}$	$\frac{19}{100}$ $=\frac{3.8}{20}$	$\frac{52}{200}$ $=\frac{5.2}{20}$	$\frac{46}{200}$ $=\frac{4.6}{20}$
Estimated number of blue eyeballs	6.4	3.8	5.2	4.6

Notice:
1) Each person got a different result;
2) The results from the 200-try experiments are closer together than the 100-try results are.

Step 3) CHOOSE the most likely estimate

It looks like the more tries you do, the more your results agree with each other. Especially from the last two results on the table, it seems likely that there are 5 blue eyeballs in the bag.

The Acid Test

Design an experiment to find out how likely it is that your toast and marmalade will land marmalade-side down if you drop it on the carpet.

First Quadrant Coordinates

Plotting Points

The first thing you need to know about graphs is how to plot points on a grid like this one.

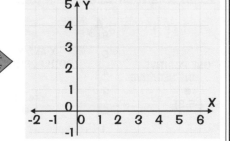

A point has two numbers to identify its position: its *COORDINATES*.

The coordinates of the points opposite are:

A(1,1)	C(4,3)
B(2,3)	D(3,1)

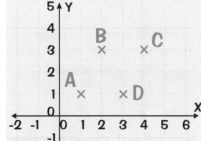

Coordinates — *Getting Them in the Right Order*

Always give COORDINATES IN BRACKETS like this: (X , Y)
Make sure you get them the right way round —

　　　　here are 3 handy rules to help you remember:

1) The two coordinates are always in ALPHABETICAL ORDER, X then Y.

2) X is always the flat axis going ACROSS the page. In other words
　"X is a..cross "　　Get it? — X is a "×"　　　(Hilarious isn't it)

3) You always go IN THE HOUSE (→) and then UP THE STAIRS (↑), so
　it's ALONG first and then UP,　i.e. X-coordinate first, then Y.

The Acid Test

LEARN how to get the coordinates the right way round.

Cover the page and write down what you have learnt.

Now try to do this:
Add two more points to make a square. Draw in the square. Write down the coordinates of the corners of the square.

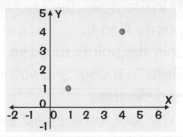

Positive and Negative Coordinates

Positive(+) and Negative(–) Coordinates

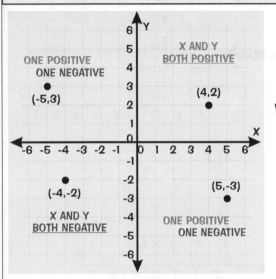

Graphs can have <u>four *different*</u> <u>*regions*</u> where the **X–** and **Y–** coordinates are either *positive* (+) or *negative* (–).

In <u>FIRST QUADRANT</u> graphs (p.65), **X–** and **Y–** coordinates are both *positive* (+).

You have to be dead careful in the other regions though, because one or both of the coordinates will be <u>negative</u>, and that always makes life difficult.

Points *on a Line*

A <u>line</u> is made up of <u>points</u>. Some of the points on the line AB are marked by ✳, these are: (2,1), (2,4), (2,-2) and (2,-3).

<u>NOTICE</u> that the **X**–coordinate for all these points is <u>2</u>. The line is called "X=2".

Some of the points on the line CD are also marked by ✳: (-2,-3), (0,-3) and (1,-3).

<u>NOTICE</u> that the **Y**–coordinate for all the points on the line is <u>-3</u>. So the line is called "Y=-3".

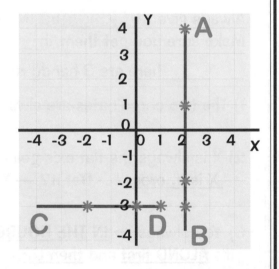

The Acid Test

1) a) Write down the coordinates of all the points A to E.
b) Join the points to make a house.
c) Draw in a door and write down it's coordinates.
d) What are the lines AB and AE called?

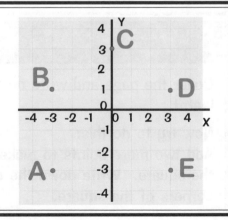

Four Graphs and a Few Words

Four Graphs To Learn

Unless you <u>LEARN</u> these graphs you won't be able to draw them in the Exam and they may well ask you to do just that.

1) _"X = a"_

A _Vertical_ Line

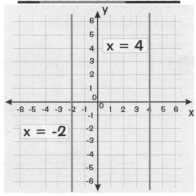

"<u>X = a number</u>" is a line that goes <u>straight up through that number</u> on the X-axis.   E.g. X = 4 goes straight up through 4 on the X-axis as shown.

2) _"Y = a"_

A _Horizontal_ Line

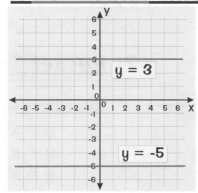

"<u>Y = a number</u>" is a line that goes <u>straight across through that number</u> on the Y-axis.   E.g. Y = -5 goes straight through -5 on the Y-axis as shown.

3) _"Y = X"_

A _Main Diagonal_

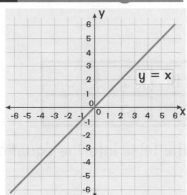

"Y = X" is the <u>main diagonal</u> that goes <u>UPHILL</u> from left to right.

4) _"Y = –X"_

The _Other Diagonal_

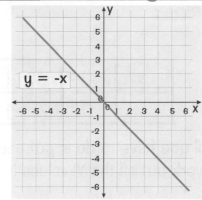

"Y = -X" is the <u>main diagonal</u> that goes <u>DOWNHILL</u> from left to right.

The Acid Test

1) <u>Learn the four graphs on this page</u>, then cover up the page and draw these graphs on a pair of X- and Y-axes:

a) Y = X   b) Y = -X   c) Y = 4   d) X = 5   e) X = -1

Drawing Graphs from Equations

There are two different types of graph you need to know:

1) Straight lines

These have equations like this:

$y = 2x$

$y = 3x + 2$

$y = x - 1$

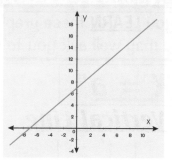

2) Bucket Shaped Graphs

Sorry, I thought you said something else.

These have equations with x^2 in:

$y = x^2$

$y = x^2 + 3$

$y = x^2 - 1$

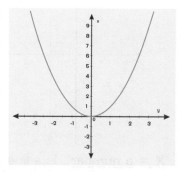

1) Doing The Table Of Values

1) What you're likely to get in the Exam is an equation (or "mapping") such as

"$Y = 2X + 1$", or "$Y = X^2$" and a half-finished table of values:

Example:

"Complete the table of values shown below using the equation $y = 2x + 1$"

x	-3	-2	-1	0	1	2	3
y	-5					5	

2) All you have to do is put each X-value from the table into the equation and work out all the Y-values in the table.

E.g. <u>For X = -3</u> $Y = 2X + 1$ $= 2 \times \text{-}3 + 1$

$= \text{-}6 + 1$ $= \underline{\text{-}5}$

This is the same as the value already given in the table, which is a good check to make sure you're doing it right

<u>For X = 1</u> $Y = 2X + 1$ $= 2 \times 1 + 1$ $= 2 + 1 = \underline{3}$

<u>For X = 3</u> $Y = 2X + 1$ $= 2 \times 3 + 1$ $= 6 + 1 = \underline{7}$

Drawing Graphs from Equations

2) Plotting The Points and Drawing The Graph

1) PLOT EACH PAIR of X- and Y- values from the table as a point on the graph.

2) Do it very CAREFULLY — and don't mix up the X- and Y-values (See P. 65/66)

3) The points will always form A DEAD STRAIGHT LINE or A COMPLETELY SMOOTH CURVE.
   NEVER let one point drag your line off in some RIDICULOUS direction:

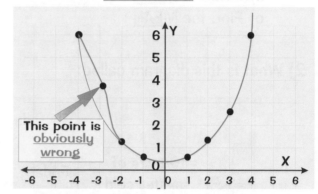

In graph plotting, you never get SPIKES or LUMPS – only MISTAKES

4) If one point does look a bit wacky then check 2 things:
   1) the y-value YOU worked out in the table and
   2) that you've plotted it properly!

Example Continued from last page:

If we carry on our Example from the last page, the completed table of values is:

x	-3	-2	-1	0	1	2	3
y	-5	-3	-1	1	3	5	7

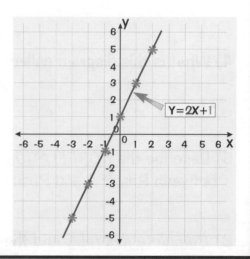

When these values are plotted and joined they make a nice STRAIGHT LINE — which crosses the y-axis at 1.

You'll also notice the line goes up 2 times as much as it goes along.

The Acid Test

x	-4	-2	-1	0	1	2	4
y	-1			3			

1) LEARN all the important details on this page.
2) Then use them to complete this table of values for the equation: Y = X + 3
3) Then plot the points on graph paper and draw the graph.

Revision Test for Section Four

WHAT YOU'RE SUPPOSED TO DO HERE is put all the methods of Section Four into practice to answer these questions.

Revision Test

1) For this set of numbers: 3, 2, 9, 5, 10, 7, 2, 6

   a) Find the <u>MODE</u>

   b) Find the <u>MEDIAN</u>

   c) Find the <u>MEAN</u>

   d) Find the <u>RANGE</u>

2) What is this diagram called:

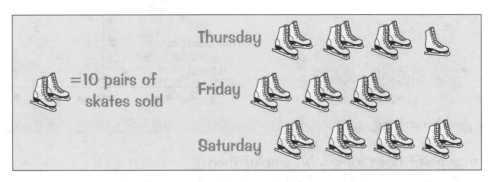

Thursday

= 10 pairs of skates sold

Friday

Saturday

3) _How many pairs of skates_ were sold on _Thursday_?

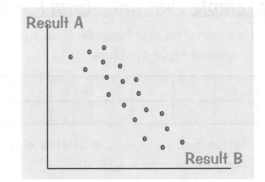

Result A

Result B

4) What is this diagram called:

5) Describe the _CORRELATION_ between Result A and Result B.

6) _Complete this table_ and then put the information into a _PIE CHART_

Fruits	Apple	Strawberry	Banana	Orange	Totals:
Number		15	10		60
Angle on Pie Chart	120		60	90	360 degrees

Revision Test for Section Four

7) In a tally table what does $20 \le w < 30$ mean? _Would you put 20 in this group_? _What about 30_, would it go in this group
or the _next one up_, $30 \le w < 40$?

8) A bag contains 6 yellow balls, 9 purple balls and 12 red balls. _Find the probability_ of picking out _a purple ball_.

9) If I toss 2 coins, _list all the possible outcomes_. Since all these outcomes are equally likely, what is the chance of getting _one head and one tail_?

10) If I toss a coin and throw a dice, _list all the possible outcomes_ and say what the probability is of me getting _a TAIL and a FIVE_.

11) The probability of a spinner landing on _BLUE_ is _0.3_.
What is the chance of this spinner _NOT_ landing on blue?

12) _Plot these points_ on the grid shown:

        A(-6,-6)
        B(6,6)
        C(-6,6)
        D(6,-6)

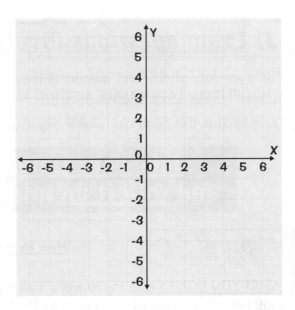

13) Join (i) A to B; (ii) C to D. What are the equations of these two lines?

SO IF YOU GET STUCK with any of these questions, have a look back at the right page in Section Four to find out how to do it!

Clocks and Calendars

Since every video recorder in the country uses the 24 hour clock you must know how to read it by now. The only thing you might need reminding about is "am" and "pm" in the 12 hour clock:

`20:23:47`

`08:23:47`

1) *am and pm*

"am" means "Morning"
"pm" means "Afternoon and Evening"

"am" runs <u>from 12 midnight to 12 noon</u>
"pm" runs <u>from 12 noon to 12 midnight</u>

(but you must know that already, surely)

2) *Conversions*

You'll definitely need to know these very important facts:

1 day = 24 hours
1 hour = 60 minutes
1 minute = 60 seconds

3) *Exam questions involving time*

There are lots of different questions they can ask involving time but the same <u>GOOD OLD RELIABLE DEPENDABLE METHOD</u> will work wonders on all of them.

"And what is this good old reliable dependable method?", I hear you cry.   Well, it's this:

Take your time, write it down, and split it up into SHORT EASY STAGES

Example 1   *"How long is it from 7.45 to 12.10?"*

<u>WHAT YOU DON'T DO</u> is try to work it all out in your head <u>in one go</u> — this ridiculous method <u>fails nearly every time</u>.   INSTEAD, DO THIS:

"Take your time, write it down, and split it up into SHORT EASY STAGES"

7.45 ⟶ 8.00 ⟶ 12.00 ⟶ 12.10
15 mins   4 hours   10 mins

This is a nice safe way of finding the total time from 7.45 to 12.10:

4 hours + 15 mins + 10 mins = <u>4 hours 25 mins</u>.

Clocks and Calendars

Example 2

"A train sets off at 1120 and arrives at its destination at 14.17. How long was the journey?"

Use just the same method: split it up into SHORT EASY STAGES.

11.20 ⟹ 12.00 ⟹ 14.00 ⟹ 14.17
40 mins   2 hours   17 mins

Add the times: 2 hours + 40 mins + 17 mins.

The train journey took 2 hours 57 mins.

4) Exam Questions Involving Calendars

1999	JANUARY	FEBRUARY	MARCH
Monday	4 11 18 25	1 8 15 22	1 8 15 22 29
Tuesday	5 12 19 26	2 9 16 23	2 9 16 23 30
Wednesday	6 13 20 27	3 10 17 24	3 10 17 24 31
Thursday	7 14 21 28	4 11 18 25	4 11 18 25
Friday	1 8 15 22 29	5 12 19 26	5 12 19 26
Saturday	2 9 16 23 30	6 13 20 27	6 13 20 27
Sunday	3 10 17 24 31	7 14 21 28	7 14 21 28

No, not colanders.

Example 1

"How many days are there from January 24th 1999 to March 9th 1999?"

ANSWER: Jan 24th ⟹ End Jan ⟹ End Feb ⟹ Mar 9th
7 days   28 days   9 days

Add these to get the total days: 28 days + 9 days + 7 days = 44 days.

Example 2

"Which day in January 1999 has the numbers from the 7 times table as its dates?"

ANSWER:

Look at January for 7, 14, 21, 28 and you find it is THURSDAY.

Example 3

"A restaurant has its two week annual holiday in January. Its first day closed is January 4th. When does it reopen?"

ANSWER:

Simple. Just count 2 weeks on the calendar from January 4th and you come to January 18th. The restaurant reopens on MONDAY, JANUARY 18th.

The Acid Test

1) In the 12 hour clock, what is a) 18.30; b) 14.45?
2) A bus sets off at 8.35 am. The journey takes 2 hrs 40 mins. When does it arrive?
3) How many days are there from January 22 1999 to March 1 1999?

Compass Directions and Bearings

The Eight Points of the Compass

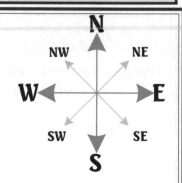

Make sure you know all these directions on the compass.

These eight compass directions and all the ones in between can be expressed as BEARINGS.

Bearings

1) A bearing is the DIRECTION TRAVELLED, GIVEN AS AN ANGLE in degrees.
2) All bearings are measured CLOCKWISE from the NORTHLINE.
3) All bearings are given as 3 figures: e.g. 060° rather than just 60°, 020° rather than 20° etc.

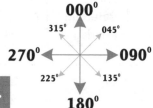

The bearing of A from B

LEARN the BEARINGS of these *compass directions*.

Points Of the Compass and Right Angles

The 4 main compass directions (N, E, S, W) are separated by right angles.

EXAMPLE 1: Turning from North to South or South to North you move through *two* right angles.

EXAMPLE 2: If you turn from North to East, or South to West, you move through *one* right angle.

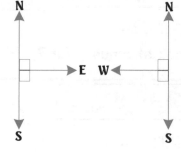

The Acid Test

1) Draw a diagram of the 4 main compass points showing the right angles.
2) A plane is flying due North. It turns clockwise on a bearing of 270°. How many right angles did it move through?
3) A man walks due South. What is his bearing?
4) What is the bearing for: a) South West; b) North West?

Compass Directions and Bearings

The 3 Key Words

Only learn this if you want to get bearings *RIGHT*

1) "FROM"

Find the word "FROM" in the question, and put your pencil on the diagram at the point you are going "*from*".

2) NORTHLINE

At the point you are going "FROM", *draw in a NORTHLINE*.

3) CLOCKWISE

Now draw in the angle CLOCKWISE *from the northline to the line joining the two points*. This angle is the BEARING.

Example

Find the bearing of S from T:

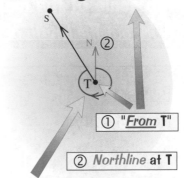

① "*From* T"

② *Northline* at T

③ Clockwise, from the N-line.

This angle is the bearing of S from T and is 330°.

The Acid Test

1) a) What is the bearing of B from A?

   b) What is the bearing of A from B? (need new northline).

2) a) What is the bearing of S from R?

   b) What is the bearing of T from S?

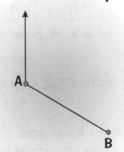

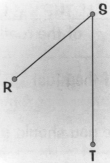

3) LEARN the 3 KEY WORDS for **BEARINGS**, then turn over and draw it out again.

Maps and Map Scales

1) The most usual map scale is "*1cm = so many km*"

2) This just tells you <u>HOW MANY km IN REAL LIFE</u> it is for <u>1cm MEASURED ON THE ACTUAL MAP ITSELF</u>.

1) *Converting* "cm on the Map" *into* "Real km"

This map shows the Grand Canal built by the Emperor of Mars in time for the 5,000,000 BC Jubilee.

The scale of the map is "1cm to 50km"
"Work out the length of the section of canal between Vombis and Ignarh."

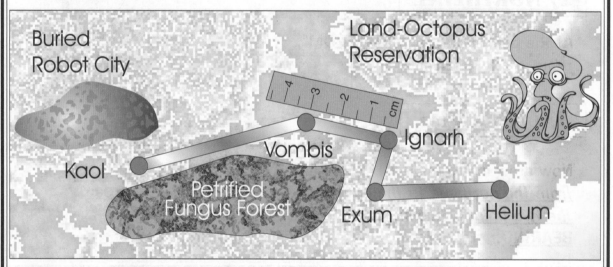

This is what you do (as shown on the diagram)

1) <u>PUT YOUR RULER AGAINST THE THING</u>
  you're finding the length of

2) <u>MARK OFF EACH WHOLE CM AND WRITE IN THE DISTANCE IN KM</u>
  next to each one

3) <u>ADD UP ALL THE KM DISTANCES TO FIND THE WHOLE LENGTH</u>
  of the road in km.   (i.e. 50km + 50km + 50km = <u>150km</u>)

Of course if they just <u>TELL YOU</u> the thing is say 4cm long *you won't be able to put your ruler on it*.

In that case you should *draw an imaginary line* 4cm long and *then mark off the km on it using your ruler just the same*, as shown in the example on the next page:

Maps and Map Scales

2) Converting "Real km" into "cm on the Map"

Example:

"A map is drawn on a scale of 1cm to 2km. If a road is 12km long in real life, how long will it be in cm on the map?"

Answer:

1) <u>Start by drawing the road as a straight line</u>:

2) <u>Mark off each cm and fill in how many km it is for each one</u>

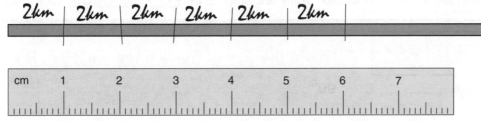

2km | 2km | 2km | 2km | 2km | 2km

cm 1 2 3 4 5 6 7

3) Keep going <u>until the km add up to the full distance</u>
(12km in this case).

Then just <u>count how many cm long your line is</u>
(In this case <u>6cm</u>).

The Acid Test

1) LEARN the 3 rules for working with map scales.

2) Estimate the length in metres of HMS Gynormous shown here:

SCALE: 1cm to 80m.

HMS GYNORMOUS

3) How many cm long would a 480m aircraft carrier be?

Lines and Angles

Angles aren't that bad — you just have to learn them, that's all!

1) Estimating Angles

The secret here is to <u>KNOW THESE FOUR SPECIAL ANGLES</u> as *reference points*. Then you can COMPARE any other angle to them.

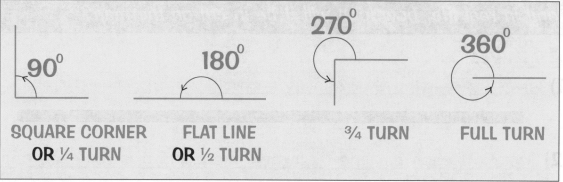

| 90^0 | 180^0 | 270^0 | 360^0 |
| SQUARE CORNER OR ¼ TURN | FLAT LINE OR ½ TURN | ¾ TURN | FULL TURN |

When two lines meet at 90° they are said to be <u>PERPENDICULAR</u> to each other.

Examples:

Estimate the size of these three angles A, B and C:

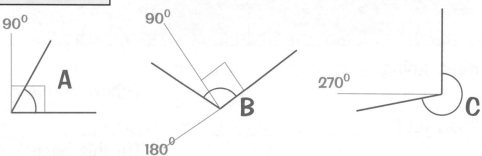

If you *compare each angle* to the <u>reference angles of 90°, 180° and 270°</u> you can easily estimate that:

<u>A = 60°</u>, <u>B = 110°</u>, <u>C = 260°</u>

The Acid Test

LEARN the four main reference angles.

1) Estimate these angles:

a)   b)   c)   d)

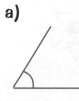

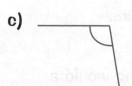
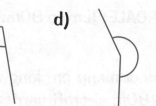

Measuring Angles with Protractors

The _2 big mistakes_ that people make with PROTRACTORS:

> 1) Not putting the 0° line at the start position
>
> 2) Reading from the WRONG SCALE.

Two Rules for Getting it right!

1) ALWAYS position the protractor with the _bottom edge_ of it along one of the lines as shown here:

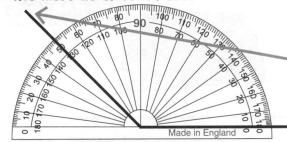

Count in 10° steps from the _start line_ right round to the _other one_ over there.

Nice tractor Ted.   _Thanks Bill._   PRO

← _Start line_

2) COUNT THE ANGLE IN 10° STEPS

from the start line right round to the other one.

> _DON'T JUST READ A NUMBER OFF THE SCALE_ – it will probably be the
> WRONG ONE _because there are_ TWO scales to choose from.
> _The answer here is 130° – NOT 50°! – which you will only get right if you start
> counting 10°, 20°, 30°, 40° etc. from the start line until you reach the other line.
> You should also_ estimate _it as a check._

Acute **Angles**	_Obtuse **Angles**_	_Right **Angles**_
SHARP POINTY ONES (less than 90°)	FLATTER-LOOKING ONES (between 90° and 180°)	SQUARE CORNERS (exactly 90°)

The Acid Test

1) LEARN the 2 rules for using protractors.

2) LEARN what ACUTE, OBTUSE and RIGHT ANGLES are. Draw one example of each.

3) Measure and draw these angles: a) 55°   b) 120°   c) 170°

Five Angle Rules

1) ANGLES IN A TRIANGLE

Add up to 180°.

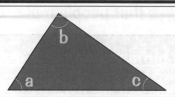

$a+b+c=180°$

2) ANGLES ON A STRAIGHT LINE

Add up to 180°.

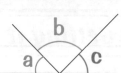

$a+b+c=180°$

3) ANGLES IN A 4-SIDED SHAPE
(a "Quadrilateral")

Add up to 360°.

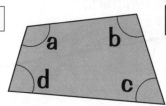

$a+b+c+d=360°$

4) ANGLES ROUND A POINT

Add up to 360°.

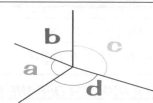

$a+b+c+d=360°$

5) ISOSCELES TRIANGLES

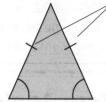

These dashes indicate two sides the same length

**2 sides the same
2 angles the same**

In an isosceles triangle, UNDERLINE YOU ONLY NEED TO KNOW ONE ANGLE to be able to find the other two, which is *very useful* IF YOU REMEMBER IT.

1)

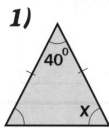

$180° - 40° = 140°$
The two bottom angles are both the same and so they must add up to 140°, so each one must be half of 140° (= 70°). So X = 70°

2)

The two bottom angles must be the same, so 50°+ 50°= 100°. All the angles add up to 180° so Y = 180° − 100° = 80°.

The Acid Test

1) There are **FIVE ANGLE RULES** on this page. **LEARN** them, then turn over and see how much of it you can write down.

2) Find the size of angle Z in the triangle shown:

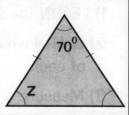

Three-letter Angle Notation

The best way to say which angle
you're talking about in a diagram is
by using <u>THREE letters</u>.

For example in the diagram,
<u>angle ACB = 25°</u>.

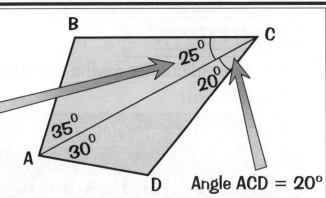

Angle ACD = 20°

In Three-letter Notation:

1) The <u>MIDDLE LETTER</u> is <u>where the angle is</u>.
2) The <u>OTHER TWO LETTERS</u> tell you <u>WHICH TWO LINES enclose the angle</u>.

Examples:

1) Angle BCD is <u>AT C</u> and is <u>ENCLOSED BY</u> *the lines* BC *and* CD
   (you just split BCD into BC-CD).   <u>Angle BCD = 45°</u>.

2) Angle ACD (AC-CD) is <u>AT C</u> and is <u>ENCLOSED BY</u> *the lines* AC
   and CD.   <u>Angle ACD = 20°</u>.

This is the way they'll do it in the exam so like it or lump it, you'd better get the
hang of it. It's very simple though, don't you think?

Interior and Exterior Angles

1) A polygon is just a fancy word for a shape with lots of sides.
2) A *regular polygon* is one where all the sides and angles are the same
                (See P.34).
3) An *irregular polygon* is one where the sides and
                angles are different like this one:

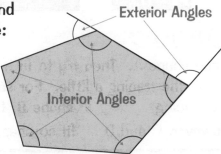

Exterior Angles

Even on irregular polygons you can have
Interior and Exterior angles. Make sure you
learn this diagram so that you know what
Interior and Exterior angles are.

Interior Angles

The Acid Test

1) Looking at the diagram at the top of the page, write down the size of angle
     CAD and also give the three-letter notation for the angles which are
          a) 35°   and   b) 65°

Congruence

Congruent

This is a ridiculous maths word which sounds really complicated when it's not:

> If two shapes are **CONGRUENT**, they are simply **THE SAME**
>
> — the **SAME SIZE** and the **SAME SHAPE**.
>
> That's all it is.   Just make sure you know the word.

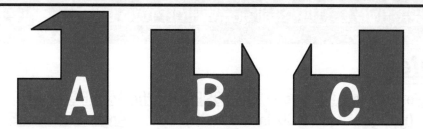

CONGRUENT: *Same* size, *same* shape

Example:

"Which shapes are congruent to A?"

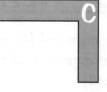

Answer:

Trace shape A. Then try to fit it onto shapes B, C, D. It fits easily onto D — just *rotate* the tracing a little. For shape C, *rotate* AND *flip over*. Whatever you do, the tracing *will not fit* shape B, that is, shape B is not congruent to shape A. However, C and D *do* fit so shapes C and D *are congruent* to A.

The Acid Test

1) Write down what congruent means.

2) Draw three shapes that are all congruent.

Reflection and Rotation

Rotation

Where a shape has been rotated, you could be asked for these 3 details, so best learn them.

1) **ANGLE** turned
2) **DIRECTION** (Clockwise or Anti-clockwise)
3) **CENTRE** of Rotation

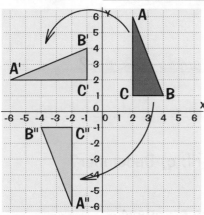

ABC to A'B'C' is a Rotation of _90°, anticlockwise_, **ABOUT** _the origin_.

ABC to A"B"C" is a Rotation of _half a turn (180°), clockwise_, **ABOUT** _the origin_.

Reflection

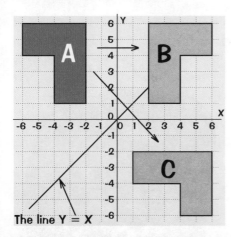

The line Y = X

Where a shape has been reflected, you must give this one detail:

1) The **MIRROR LINE**

A to B is a _reflection IN the Y-axis_.

A to C is a _reflection IN the line Y=X_.

The Acid Test

1) Copy the diagram onto squared paper.
2) Draw a reflection of A (A') with the x-axis as the mirror line.
3) Draw another reflection of A (A"); this time use the y-axis as the mirror line.
4) Describe the transformation A →B.
5) Describe the transformations A' →B, A" →B.

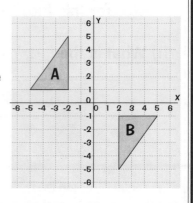

Revision Test for Section Five

WHAT YOU'RE SUPPOSED TO DO HERE is put all the methods of Section Five into practise to answer these questions.

Revision Test

1) What is *2020* in *12 hour clock*?

2) What is *7.30pm* in *24 hour clock*?

3) How many *hours and minutes* is it from *10.50am to 4.40pm*?

4) A bus sets off at *9.25am*. Its journey takes *5 hours and 20 minutes*. At *what time* does it arrive at its destination?

5) How many days are there in a) January; b) April; c) December?

6) Draw a diagram showing the *eight points* of the compass.

7) How many right angles is a) 180°; b) 270°?

8) What are the *Three Key Words* for *Bearings*?

9) A man sets off on a long walk from village Z to go *on a bearing of 075°*. *Draw a line* on the picture to show the bearing.

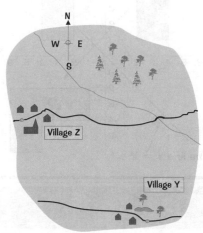

10) A group of school children also leave village Z to go for a walk. The direction they take is a *bearing of 150°*. *Draw a line* for this too.

11) What is the *bearing* of Village Z *from* Village Y?

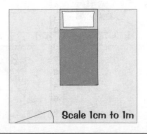
Scale 1cm to 1m

12) This is a scale drawing of a bedroom. The scale is 1cm to 1m.
   a) What are the *actual dimensions* of the *room*?
   b) What are the *dimensions of the bed*?

SECTION FIVE — ANGLES AND OTHER BITS

Revision Test for Section Five

13) Using a _scale of 1cm to 8km_, how many
    kilometres does this line represents?

14) Draw these _four special angles_:
        a) 90°   b) 180°   c) 270°   d) 360°

15) <u>ESTIMATE</u> these angles and then <u>MEASURE</u> them, and make sure
    your two answers are similar for each angle:
      a)   b)   c)   d)

16) What are   a) _Acute angles_
                          b) _Obtuse angles_
                          c) _Right-angles_?

17) Work out _angles X and Y_ in the diagram: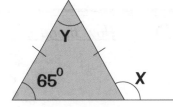

18) If two shapes are _the same size_ and _same shape_ what is the _word_ used
    to describe them?

19) a) What are the _3 details for rotation_?
        b) What is the _one detail for reflection_?

SO IF YOU GET STUCK with any of these questions, _have a look
back at the right page_ in Section Five to _find out how to do it_!

Balancing

This is a favourite topic with examiners — not surprisingly since it naturally leads on to the dreaded subject of EQUATIONS.

Balancing and EQUATIONS

In fact, "balancing" already _is_ "equations", and there's no catch — it really is _as easy as it looks_.

Example

First Fact

Two copper bracelets   One silver dish

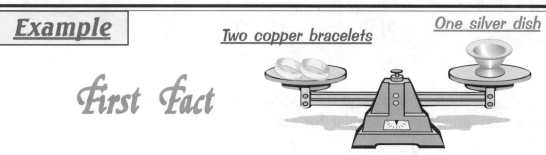

BALANCE

AND

Two silver dishes   One gold cup

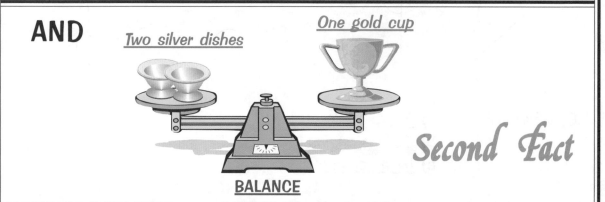

Second Fact

BALANCE

SO

How many copper bracelets   One gold cup

?

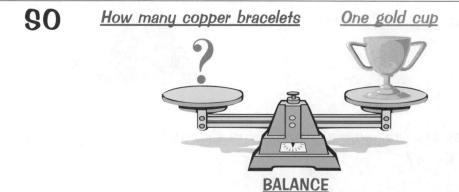

BALANCE

The two facts tell you that one gold cup balances with 4 copper bracelets. Now take a step back and:

Ask yourself _HOW YOU DID IT_, because that's what's important — _THE METHOD_

Balancing

Method

The technical term for what you've just done is "substitution", but to save ink we'll just call it _replacing_.

Step 1) What you did was to look back at the _First Fact_ and see that one silver dish weighed the same as two copper bracelets

Step 2) _Taking this fact absolutely seriously_, you figured that _ANY TIME_ you saw a silver dish, you could replace it with two copper bracelets.

SO, INSTEAD OF

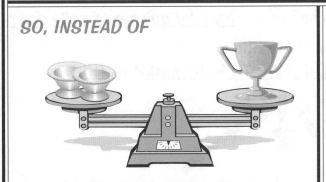

YOU GET

The Acid Test

Use the __REPLACING__ method (also known as the _taking-the-information-you've-been-given-seriously_ method) to fill in what should go in place of these question marks:

1) How many copper bracelets

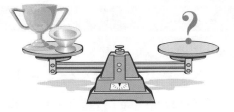

2) How many copper bracelets

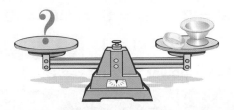

3) How many silver dishes

4) How many copper bracelets

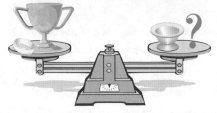

Powers

Not **Complicated**

1) People think with a name like "powers" (or "indices") these things must be really complicated   <u>BUT THEY'RE NOT</u>.

2) They're just a _simple shorthand_ for something very ordinary:

$$\underline{6 \times 6 \times 6 \times 6 \times 6 = 6^5}$$

1) _Somebody, somewhere decided (in their infinite wisdom)_ that instead of writing say, $6 \times 6 \times 6 \times 6 \times 6$, it would be a lot better to just write it as: 6^5

2) So 6^5 just means "_6 times by itself 5 times_",   (nothing too tricky there)

3) And let's face it, you can see the advantage—_it's a lot quicker to write down_

4) But as well as being quicker, it's not that difficult is it?

— <u>BECAUSE ALL YOU HAVE TO REMEMBER IS THIS</u>:

Powers are ace.

$$4^6$$

THIS NUMBER   .....<u>_times by itself_</u> ....THIS MANY TIMES

$$4^6 = 4 \times 4 \times 4 \times 4 \times 4 \times 4 = \underline{4096}$$

Six Important **Examples**

1) How much is 2^7 ("two to the power 7")?   ANS: $2 \times 2 \times 2 \times 2 \times 2 \times 2 \times 2 = \underline{128}$

2) What is 7^2 ("7 squared")? 7^2 is $7 \times 7 = \underline{49}$

3) What is 4^3 ("four cubed")? 4^3 is $4 \times 4 \times 4 = \underline{64}$ "The cube of 4 is 64"

4) $3^1 = 3$ $5^1 = 5$ $10^1 = 10$ (anything to the power 1 is just itself)

5) $1^4 = 1 \times 1 \times 1 \times 1 = 1$ — in fact 1 to any power is always just 1

6) What power of 2 makes 32 ?   Try it out: $2 \times 2 \times 2 = 8$, $2 \times 2 \times 2 \times 2 = 16$
$2 \times 2 \times 2 \times 2 \times 2 = 32$, so 32 is 2^5 (because <u>TWO</u> times by itself <u>FIVE</u> times = 32)

The Acid Test

TURN OVER THE PAGE and <u>write down what POWERS are, with an example.</u>

1) Find the value of 2^6.
2) What is the value of "5 squared"? What is the cube of 6?
3) What is the value of x if $7^x = 343$ (i.e what power of 7 makes 343)?
4) What is the value of $3^4 \times 2^1 \times 1^8$?

Square Roots

A bit Complicated but not that difficult

Square roots aren't too bad so long as you know *The big Secret* which is to ALWAYS SEE THEM IN REVERSE.

Square Root —Turn it Round

EXAMPLE: "Find the SQUARE ROOT of 49"

In reverse: "What number TIMES BY ITSELF gives 49?"

You should now see that *the answer is 7*, because "7 times by itself = 49"

> ## "Square Root" means "What Number Times By Itself..."

The special symbol for square root is $\sqrt{}$ so $\sqrt{49}$ means "square root of 49".

Calculator Buttons for Powers and Roots

The Square Root Button $\boxed{\sqrt{}}$

The $\boxed{\sqrt{}}$ button gives the *square root* of the number in the display.

Try this: $\boxed{25}$ $\boxed{\sqrt{}}$ .... you should get 5

5 is known as THE SQUARE ROOT of 25 because $5 \times 5 = 25$.

The Powers Button: $\boxed{x^y}$

On most calculators this is the *2nd function of the* $\boxed{\times}$ *button*. It is used for working out powers of numbers quickly.

For example to find 3^5:
Instead of pressing $3 \times 3 \times 3 \times 3 \times 3$ you should just press

$\boxed{3}$ $\boxed{x^y}$ $\boxed{5}$ $\boxed{=}$ (or $\boxed{3}$ $\boxed{\text{SHIFT}}$ $\boxed{\times}$ $\boxed{5}$ $\boxed{=}$)

......which gives 243.

The Acid Test

LEARN what it says in the 2 shaded boxes above, then COVER UP THE PAGE and write it down.

1) Find the square root of 64.
2) Find   a) $\sqrt{36}$; b) $\sqrt{40}$; c) 3^9;   d) 2^8.
3) The square shown has an area of 16m². Find X?

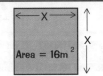

Area = 16m²

Number Patterns and Sequences

There are five different types of number sequences they could give you. They're not difficult — AS LONG AS YOU WRITE WHAT'S HAPPENING IN EACH GAP.

1) "Add or Subtract the Same Number"

The SECRET is to _write the differences in the gaps_ between each pair of numbers:

E.g.   3   8   13   18   ...   23   20   17   14   11   ...
           +5   +5   +5   +5   -3   -3   -3   -3   -3

The RULE: "Add 5 to the _previous term_"   "Subtract 3 from the _previous term_"

2) "Add or Subtract a Changing Number"

Again, WRITE THE CHANGE IN THE GAPS, as shown here:

E.g.   2   4   7   11   16   ...   or   30   23   17   12   8   ...
          +2   +3   +4   +5   +6   -7   -6   -5   -4   -3

The RULE: "Add 1 _extra_ each time to the _previous term_"   "Subtract 1 _extra_ from the _previous term_"

3) Multiply by the Same Number each Time

This type have a common MULTIPLIER linking each pair of numbers:

E.g.   2   4   8   16   ...   4   12   36   108   ...
              ×2   ×2   ×2   ×2   ×3   ×3   ×3   ×3

The RULE: "Multiply the _previous term_ by 2"   Multiply the _previous term_ by 3"

4) Divide by the Same Number each Time

This type have the same DIVIDER between each pair of numbers:

E.g.   400   200   100   50   ...   70 000   7000   700   70   ...
            ÷2   ÷2   ÷2   ÷2   ÷10   ÷10   ÷10   ÷10

The RULE: "Divide the _previous term_ by 2"   "Divide the _previous term_ by 10"

5) Add the Previous Two Terms

This type of sequence works by adding the last two numbers to get the next one.

E.g.   1   1   2   3   5   8   13   ...   2   4   6   10   16   26 ..
              1+1 1+2 2+3 3+5 5+8 8+13   2+4   4+6 6+10 10+16

The RULE: "Add the previous two terms"

Number Patterns: A Typical Question

What we're going to look at is a ...

"State the rule for extending the pattern"

... type of question.

This is what a lot of _Exam questions_ end up asking for and it's easy enough _so long as you remember this:_

ALWAYS say what you do to the **PREVIOUS TERM** to get the next term.

All the number sequences on the opposite page have the rule for extending the pattern written in the box underneath them. Notice that they all refer to the _previous term_.

BUT: — You may not always be given a number pattern as a string of numbers. In fact, it's highly likely that they'll start by giving you a _series of picture patterns_ instead.

Example

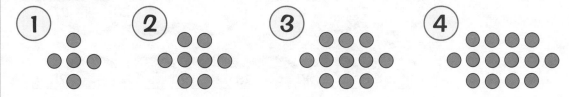

You might get asked, _"How many dots are in pattern number ⑤ ?"_

Method

Just turn it into a _number_ sequence and you'll get the answer soon enough:

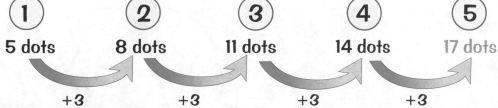

The Acid Test

1) Write a number sequence for the _blue_ dots only in the above series of patterns.
2) How many blue dots will there be when there are:
   a) 5 red dots;   b) 6 red dots;   c) 9 red dots?

Number Patterns Formula

Finding the nth number

An Exam question might ask you to *"give an expression for the nth number in the sequence."*

Let's say you're faced with this sequence:

3, 7, 11, 15, 19, 23, 27   and you're asked to *"Find the nth number"*.

First of all , WHAT does the question MEAN? It just means, *"Find the aNythingeth number"*. In other words,

> A rule that will work for finding the four*th*, fif*th*, six*th*, million*th* or ...*nth* number in the sequence.

For this type of Exam question, you have a CHOICE of TWO APPROACHES.

1) Using Times Tables

This involves, comparing the sequence with the NEAREST TIMES TABLE:

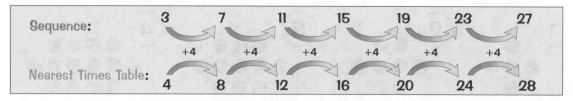

What's important is the *gap* between each number.

That's what shows you the "family" it belongs to — in this case, the *Four Times Table* family.

In fact, the sequence we started with is just like the Four Times Table except that it's shifted 1 number to the left:

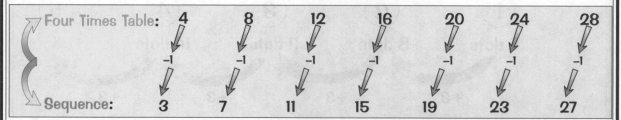

So, the 20th number in the sequence, for example,

will just be 20 fours less 1   = $20 \times 4 - 1$

   = $80 - 1$

   =   79

and the nth number will just be n fours less 1   = $n \times 4 - 1$

   = $4n - 1$

Number Patterns Formula

2) Using the Formula:

Using this approach, you can work out the answer more or less in your sleep provided that you *LEARN THE FORMULA*:

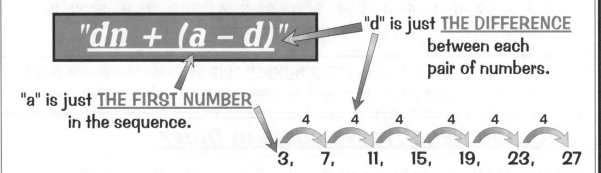

"dn + (a – d)"

"d" is just THE DIFFERENCE between each pair of numbers.

"a" is just THE FIRST NUMBER in the sequence.

4 4 4 4 4 4

3, 7, 11, 15, 19, 23, 27

To get the *nth term*, you just *find the values of "a" and "d" from the sequence and stick them in the formula*.

You don't replace n though — that wants to stay as n

— **of course YOU HAVE TO LEARN THE FORMULA, but life is like that.**

Example:

"Find the nth number of this sequence:

5, 8, 11, 14 ..."

ANSWER:

1) The formula is dn + (a – d)

2) The first number is 5, so $\underline{a = 5}$.   The differences are 3 so $\underline{d = 3}$.

3) Putting these in the formula gives: 3n + (5 – 3) = 3n + 2.

So the nth number for this sequence is given by: "3n + 2".

The Acid Test

LEARN the 5 types of number patterns and the formula for finding the nth number.

1) Find the next two numbers in each of these sequences, and say in words what the rule is for extending each one:
   a) 6, 13, 20, 27 ....   b) 8, 80, 800 ....   c) 128, 64, 32, 16, 8 ...

2) Find the expression for the nth number in this sequence: 5, 7, 9, 11 ...

3) Two tables for four, when pushed together, seat 6. How many can be seated if :
   a) 3   and b) n   such tables are pushed together?

Negative Numbers

THE NUMBER LINE

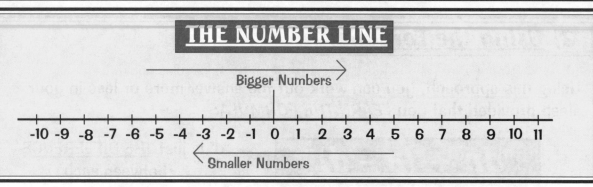

You need to *remember* the diagram of <u>THE NUMBER LINE</u> as shown above — it could be the answer to all your problems — well, all your *negative number problems* anyway.

1) *Putting Negative Numbers in Order*

<u>EXAMPLE</u>: <u>Put these in order of size</u>: **6,   -9,   2,   -5,   -2,   11,   -7,   8**

ANS: *1) Quickly <u>draw out the full Number Line</u> as shown below*
               2) Put the numbers <u>in the same order as they appear on the number line</u>.

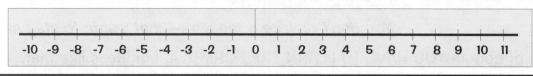

-9	-7	-5	-2	2	6	8	11

So in order of size they are: <u>-9, -7, -5, -2, 2, 6, 8, 11</u>

> <u>Note that</u> -5 is <u>BIGGER</u> than -7, because it is <u>FURTHER UP THE NUMBER LINE</u>.
> <u>Negative numbers go the "wrong way"</u> — smaller numbers are bigger!

2) *Finding The Range of Values*

A very common Exam question is for the <u>RANGE OF TEMPERATURE</u> for a place where it goes below freezing at night.

> <u>EXAMPLE</u>: One day the temperatures in Moscow were: Midday —   12°C
>   Midnight —   -10°C
>   What was the full <u>RANGE</u> of temperature?

<u>ANSWER</u>: Once again, just do a *quick sketch of the Number Line*, mark the two temperatures on it and then just *<u>count how many degrees it is between them</u>* — easy:

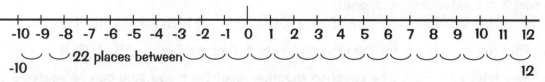

The answer is: The full range of temperature in Moscow was <u>22°C</u>.

Section Six — Algebra

Negative Numbers

1) Multiplying and Dividing

For *multiplying and dividing with* NEGATIVE NUMBERS these rules apply.

+	+	makes	+
+	–	makes	–
–	+	makes	–
–	–	makes	+

Drrrrrrrrrr...

This is the best way to divide...

E.g $-2 \times 3 = -6$ (– and + makes – ,   so you get -6, not +6)

              $5 \times -4 = -20$ (+ and – makes –, so you get –20, not +20)

            $-12 \div 2 = -6$ (– and + makes –,   so you get –6, not +6)

         $-6 \times -8 = +48$ (– and – makes +,   so you get +48, not –48)

2) IF POSSIBLE*, Use Your Calculator To Do*

Negative Numbers

This is definitely the easiest way to do negative numbers — you just have to learn one thing – HOW TO PUT A NEGATIVE NUMBER INTO THE CALCULATOR:

Press the ⊟ Button AFTER Entering The Number

If you can remember that, the rest is easy.

FIVE IMPORTANT EXAMPLES:

1) Enter the number -20 into the calculator. ANSWER: Press [20] [+/–] which gives -20

2) Find -5 × -4.   ANSWER: Press [5] [+/–] [×] [4] [+/–] [=] and the answer is 20.

3) If Y = X² , find Y when X = -4.
    ANSWER: -4^2 is -4×-4 so press [4] [+/–] [×] [4] [+/–] [=] to get 16

4) If Y = 2X + 4, find Y when X = -7.
    ANSWER: Y = 2 × -7 + 4 so press [2] [×] [7] [+/–] [=] [+] [4] [=] = -10

5) To work out -3 + -5 – -2, press [3] [+/–] [+] [5] [+/–] [=] [–] [2] [+/–] [=] to get -6

The Acid Test

1) Arrange these in order of size: -10, 6, -11, -1, 5, -4, 20, -21, 22, 0

2) One day the temperature went from -4°C to 9°C. What rise in temperature was this?

3) Work these two out without your calculator: a) 5 × -10   b) -6 ÷ -2.

4) If Y = X² use your calculator to find Y when X = -2 and also when X = -3.

5) Use you calculator to work out -17 + 3 – -2.

Basic Algebra

Algebra is a pretty grim subject at the best of times but these two bits of algebra aren't too bad and I'd say they're worth having a go at — and let's face it, who are you to argue.

1) Simplifying or "Collecting Like Terms"

EXAMPLE: "Simplify 3x − 2 + 4x + 5"

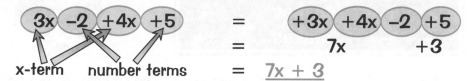

$$3x \quad -2 \quad +4x \quad +5 \quad = \quad +3x \quad +4x \quad -2 \quad +5$$
$$= \quad 7x \qquad +3$$
$$= \quad 7x + 3$$

x-term   number terms

1) Put bubbles round each term, — be sure you *capture the +/− sign* IN FRONT *of each term*.

2) Then you can *move the bubbles into the best order* so that LIKE TERMS *are together*.

3) " LIKE TERMS" have exactly the same letters, e.g. x-terms or y-terms, or no letters at all if they're number-only terms.

4) Combine LIKE TERMS using the NUMBER LINE (not the other rule for negative numbers).

2) Multiplying out Brackets

1) The thing UNDERSIDE the brackets multiplies each separate term INSIDE the brackets

2) When letters are multiplied together, they are just written next to each other, like this: p×q = pq

3) Remember, R x R = R², and TY² means TxYxY

Examples:

1) 5(2x + 3) = 10x + 15   2) 3(4y + x − 2) = 12y + 3x − 6

3) 2t(5r − 6w) = 10tr − 12tw   4) p(1 + 4q − 2r) = p + 4pq − 2rp

The Acid Test

1) Simplify these expressions: a) 6x + 7 − x + 1;   b) 9y - 4z - 2y + 3z.

2) Multiply out these expressions: a) 4g(2 + 3h − 3);   b) 7(5d − 2 − 4f²).

°F and °C Temperature Formulas

Degrees Fahrenheit (°F) is the old-fashioned temperature scale
Degrees Celsius (°C) is the more modern temperature scale

Both scales are still used a lot so make sure you know these 3 IMPORTANT DAY-TO-DAY TEMPERATURES:

Freezing Point	Room Temp.	Very Hot Day
0°C	20°C	30°C
32°F	70°F	90°F

It's mighty useful if you know these 3 temperatures in both °F and °C so that when they ask you to work one out in the Exam you can check if your answer is sensible.

For example if they asked you to convert 10°C to °F then you would expect an answer somewhere between freezing point and room temperature which would be between 32°F and 70°F, perhaps 50°F. So if you worked it out and your answer was say 80°F, YOU'D KNOW IT WAS WRONG, wouldn't you!

The Two Conversion Formulas

These two formulas come up quite a lot in Exams so it's a good thing if you're already familiar with them, but you don't have to learn them:

$$F = \frac{9}{5}C + 32$$

They're a lot easier than they look too — as long as you do them step by step.

$$C = \frac{5}{9}(F - 32)$$

EXAMPLE: "Convert a temperature of 68°F into °C"

ANSWER: (Step by step Method)

Step 1: $C = \frac{5}{9}(F - 32)$ ⟵   Write down the formula

Step 2: $C = \frac{5}{9}(68 - 32)$ ⟵   Fill in the value for F (=68)

Step 3: $C = \frac{5}{9}(36)$ so $C = 5 \div 9 \times 36 = 20$ ⟵   Work it out IN STAGES

So 68°F converts to 20°C   (seems sensible – both are a room temperature)

The Acid Test

1) Using the formula $F = \frac{9}{5}C + 32$, convert 25°C into °F. Is that a hot day or a cold day?

2) Using the other formula, find the temperature in °C equivalent to 41°F. Will that be hot weather or cold weather?

SECTION SIX — ALGEBRA

Making Formulas From Words

These can seem a bit confusing but they're not as bad as they look once you know the "tricks of the trade" as it were. There are two main types.

Type 1

(See also pages 90-91)

In this type there are _instructions about what to do with a number_ and you have to _write it as a formula_. The only things they're likely to want you to do in the formula are:

1) Multiply X   2) Divide X   3) Square X (X^2)   4) Add or subtract a number

EXAMPLE 1:   " _To find Y, multiply X by two and then subtract three_"

ANSWER: Start with X $\implies$ 2X $\implies$ <u>2X – 3</u>   so $\underline{Y = 2X - 3}$

                      Times it by 2   Subtract 3   (not too gruelling, is it?)

EXAMPLE 2:   This is the most difficult you'd ever get:

"_To find Y, square X, divide this by five and then subtract four. Write a formula for Y._"

ANSWER: Start with X $\implies$ X^2 $\implies$ $\dfrac{x^2}{5}$ $\implies$ $\dfrac{x^2}{5} - 4$

                  Square it   Divide it by 5   Subtract 4

'They're not that bad, are they? $$Y = \dfrac{x^2}{5} - 4$$

Type 2

This is a bit harder. _You have to make up a formula_ by putting in letters like "C" for "_cost_" or "n" for "_number of something-or-others_". Although it may look confusing the formulas always turn out to be **REALLY SIMPLE**, so make sure you give it a go.

EXAMPLE: Froggatt's hedgehog-flavoured potato crisps cost 35 pence a packet. Write a formula for the total cost, T, of buying n crisp-packets at 35p each.

Answer: T stands for the total cost
           n stands for the number of crisp-packets

In words the formula is: Total Cost = Number of crisp-packets × 35p

Putting the letters in: T = n ×35   or to write it better: <u>T = 35n</u>

The Acid Test

1) The value of Y is found by taking **X**, multiplying it by seven and then subtracting six. Write down a formula for Y in terms of **X**.

2) One of Froggatt's main competitors are "Ringo's", who produce a vast range of products including their widely-acclaimed "Raw Liver Spread" which costs 52p a jar. Write a formula for the total cost C pence of buying n jars of Raw Liver Spread.

Ringo's
Raw Liver
Spread

Trial and Improvement

Solving Equations

"Solving equations" just means *"Finding the value of "X" which makes the equation work"*. There are some difficult ways of doing this, but TRIAL AND ERROR is *very easy* so *make sure you learn this method*.

Example 1

Solve the equation $2X + 5 = 7 + X$ (In other words, "Find the value of X *which makes it work*")

Try X = 1

$2X + 5$	=	$7 + X$	
$2\times1 + 5$	=	$7 + 1$	
$2 + 5$	=	$7 + 1$	
7	=	8	No good (Right hand side too big)

Try X = 3

$2X + 5$	=	$7 + X$	
$2\times3 + 5$	=	$7 + 3$	
$6 + 5$	=	$7 + 3$	
11	=	10	No good (Now left hand side too big)

SO TRY IN BETWEEN:

X = 2

$2X + 5$	=	$7 + X$	
$2\times2 + 5$	=	$7 + 2$	
$4 + 5$	=	$7 + 2$	
9	=	9	Yes, it works, so $\underline{X = 2}$

Example 2

Find the value of X to one decimal place where $X^3 = 24$

ANS: Just try values of X to try and get 24. Remember, $X^3 = X\times X\times X$ (See P.92 on powers).

Try X = 4	$4^3 = 4\times4\times4 = 64$	No good, *way too big*
Try X = 2	$2^3 = 2\times2\times2 = 8$	No good, *way too small*
Try X = 3	$3^3 = 3\times3\times3 = 27$	*Very close*, just a bit too big
Try X = 2.8	$2.8^3 = 2.8\times2.8\times2.8 = 21.952$	Close again, but *too small* now
Try X = 2.9	$2.9^3 = 2.9\times2.9\times2.9 = 24.389$	That's the *nearest*

So to one decimal place, $\underline{X = 2.9}$ (ALWAYS WRITE DOWN EVERY TRIAL YOU DO)

Special Note

REMEMBER, *you don't have to think too hard with this method*. Just start with a reasonable number like 1, 2, 3 or 4, *try it in the equation*, see what answer you get and *then make a better guess* at the answer — and just keep trucking till you get it.

The Acid Test

1) Use trial and error to solve this equation: $5X - 4 = 5 + 2X$.
2) If $X^2 = 55$, use trial and error to find the value of X accurate to one decimal place.

Revision Test for Section Six

WHAT YOU'RE SUPPOSED TO DO HERE is put all the methods of Section Six into practice to answer these questions.

Revision Test

1) *Three tins* weigh the same as *one bottle*. *Two bottles* weigh the same as *one vase*. *How many tins* weigh the same as *one vase*?

2) *Work out* the value of   a) 4^3 b) 6^5 c) 8 squared
                         d) 12^4 e) $5^3 \times 4^2$ f) 7 cubed

3) *Find X* if $2^x = 128$ (Use trial and error)

4) Find to the *nearest whole number*, the <u>SQUARE ROOT</u> of
         a) 59   b) 45   c) 97

5) *Write an expression* to mean "Seven times an unknown number, less five".

6) Find the value of $\sqrt{196}$.

7) The square shown has an *area* of *81cm²*. Find the length of each side.

Area = 81cm²

8) Find: a) *next two terms*, b) the *rule for extending the pattern*, in these sequences of numbers:
     i) 2, 8, 14, 20, ...   iv) 880, 440, 220, ...
     ii) 2, 6, 11, 17, ...   v) 45, 37, 30, 24, ...
     iii) 2, 6, 18, 54, ...

9) Work out the expression for the *nth number* in this sequence:
                            7, 10, 13, 16, ...

10) On the <u>NUMBER LINE</u>, put these numbers *in order of size*, smallest first:
         -5,   8,   0,   -9,   12,   1,   -1,   2

11) One day it goes up from *-8°C to 3°C*. What is the *rise in temperature* in °C?

12) *Work out:*   a) -7 × -2 ;   b) -8 × 2 ;
                      c) 5 × -6 ;   d) -3 ÷ -1 .

Revision Test for Section Six

13) *Simplify* the expression: x + 3y + 2x – 2y – 7.

14) *Expand* this expression: 5(2m + 3n – 4).

15) *Expand* this expression: 4g(2 + 5h – 6m).

16) If G = 4HL + 6   *Find G* when H = 7 and L = 2.

17) What are the *two scales* for *measuring temperature*?

18) What is *room temperature* in ^{0}F?   What is it in ^{0}C?

19) Using the formula $F = \frac{9}{5}C + 32$, find the temperature in ^{0}F when it is
    35^{0}C. Would that temperature be a *hot day*, a *warm day* or a *cold day*?

20) "*To find N you Double M and add 3*."   Write this as a formula.

21) *Write a formula* for the cost C pence of buying n packets of "Tortoise"
    flavoured Crunchies at 27p each.

22) If the price of these crunchies is unknown and called "P", *write a formula*
    for the total cost, C of buying n packets at a price P.

23) Use *Trial and Improvement* to "solve" these equations
        (i.e. find the value of **X** that makes them work).
            a) 5X + 2 = 30 – 2X
            b) 2X + 7 = 7X – 3

24) If $X^2 = 60$, find **X** accurate to *one decimal place*.
    (Trial and Improvement)

25) If $Z^3 = 24$, find **Z** accurate to *one decimal place*.

The end of the book —
time for a monster party!

26) *How many dots* will there be in:
    a) the *fourth* pattern;
    b) the *fifth* pattern;
    c) the *twentieth* pattern?

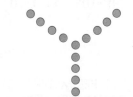

SO IF YOU GET STUCK with any of these questions, *have a look
back at the right page* in Section Six to *find out how to do it*!

Section Six — Algebra

Answers

Section One — Acid Tests

P.1 <u>BIG NUMBERS:</u> 1)a) One million, four hundred and thirty one thousand, seven hundred and sixteen.   b) Twenty five thousand, nine hundred and ninety nine.   c) Six thousand, eight hundred and twelve.   d) Two thousand and forty one.   e) One thousand, eight hundred and one.   2) 9,655. 3) 8, 26, 59, 102, 261, 3785, 4600.

P.2 <u>PLUS, MINUS, TIMES AND DIVIDE:</u> 1) 76; 76-49=27.   2) 29; 29+36=65. 3) 638; 638-392=246. 4) 358; 358+252=610. 5) 70; 70÷5=14. 6) 5; 5×20=100. 7) 952; 952÷28=34. 8) 16; 16×15=240.

P.3 <u>PATTERNS WITH TIMES AND DIVIDE:</u> 1) 4. 2) 12. 3) 8. 4) 3.   5) 6. 6) 4.

P.4 <u>MULTIPLYING BY 10, 100, 1000:</u> 1a) 1400 b) 871 c) 25000   2a) 600 b) 660 c) 21000.

P.5 <u>DIVIDING BY 10, 100, 1000:</u> 1a) 5.6 b) 4.265 c) 0.01275 2a) 2.2 b) 22.2 c) 40.

P.6 <u>MULTIPLES AND FACTORS:</u> 1) a) 4, 8, 12, 16, 20, 24, 28, 32, 36, 40, 44, 48, 52, 56, 60, 64, 68, 72, 76, 80, 84, 88, 92, 96, 100.   b) 9, 18, 27, 36, 45, 54, 63, 72, 81, 90, 99.   c) 36
2a) 1, 2, 3, 6 b) 1, 3, 5, 15.   c) 1 or 3.

P.7 <u>ODD, EVEN, SQUARE AND CUBE NUMBERS:</u> 2a) 50, 100, 132.   b) 27, 31, 49, 81, 125.   c) 49, 81, 100.   d) 27, 125.

P.9 <u>PRIME NUMBERS:</u> 1) 83, 89, 97

P.10 <u>RATIO IN THE HOME:</u> 1) 56p   2) £525:£735.

P.11 <u>RAT 'N' TOAD PIE:</u> Recipe for 8:   8 rats,   4 toads,   16 Oz of "Froggatt's Hot Sickly Sauce",   24 'taties,   A <u>very big</u> wodge of pastry (2 times as big in fact).

P.12 <u>MONEY:</u> 1) £7.57. 2) £16.45. 3) £20.97. 4) £3.05.

P.13 <u>THE BEST BUY:</u> Large size is best value at 3.7g per penny.

P.14 <u>TIMES AND DIVIDE WITHOUT A CALCULATOR:</u> 1) 336. 2) 616. 3) 832.   4) 12.   5) 121.   6) 12.

P.15 <u>TYPICAL QUESTIONS WITH × AND ÷:</u> 1) £7.45. 2) 10.

P.17 <u>CALCULATOR BUTTONS:</u> 1) See P.16   2) See P.17   3) $\boxed{5}\ \boxed{\pm}\ \boxed{\times}\ \boxed{3}\ \boxed{\pm}\ \boxed{=}$
4) 9.16×10^{14} 5) £3.10

P.18 <u>USING FORMULAS:</u> 2) 18.   3) -4.

P.19 <u>ORDERING DECIMALS:</u> 0.00049, 0.006, 0.0591, 0.082, 0.792, 1.03.

Revision Test for Section One

1) Four million, two hundred and sixteen thousand, three hundred and eighty six.   2) 4, 26, 48, 144, 612, 842, 1212, 2006. 3) a) 2680 b) 340,000   c) 0.648 d) 600 e) 200 4) Multiples are a number's times table;   10, 20, 30, 40, 50, 60 ;   3, 6, 9, 12, 15, 18;   5) Factors are what divide into a number ; 1, 2, 3, 4, 6, 8, 12, 24. 6) a) 1, 9, 16, 25, 36, 100.   b) 16, 18, 36, 100. c) 1, 9, 25, 63.
7) See P.7 2, 4, 6, 8, 10, 12, 14, 16, 18, 20 8) See P.7 1, 3, 5, 7, 9, 11, 13, 15, 17, 19
9) See P.7. 10) See P.8 2, 3, 5, 7, 11, 13 11) They must end in 1, 3, 7, 9, and they must not divide by 3 or 7;   5, 7, 11, 13, 17, 19, 23, 29. 12) Golden Rule: "DIVIDE FOR ONE THEN TIMES FOR ALL"
13) £1.90.   14) 300g. 15) a) £7.25. b) £47.94. c) £122.50.   16) Golden Rule: "DIVIDE BY THE PRICE IN PENCE" ;   Expect biggest to be best buy which it is, at 20.8g per penny.   17)
18) See P.16. 19) See P.16. 20) Just press $\boxed{\div}$ and carry on. 21) See P.17
22) $\boxed{6}\ \boxed{\pm}$ 23) 30. 24) 2.4×10¹¹ 25) £5.32. 26) See P.18. 27) 14. 28) 3.
29) 0.05, 0.15, 0.5, 0.505, 0.51, 0.55.

17) $\begin{array}{r} 43 \\ \times 28 \\ \hline 344 \\ 860 \\ \hline 1204 \end{array}$ $\begin{array}{r} 43 \\ 28\overline{)120^84} \end{array}$

Section Two — Acid Tests

P.22 <u>PERIMETERS:</u> 2)   26cm

P.25 <u>AREAS - TYPICAL QUESTIONS:</u> 1) 15cm² 2) 12cm² 3) A=L×W; 18=6×?; so ?=3 cm. 4) A=L×W; 24=?×3; so ?=8 cm.

P.27 <u>EXTRA CIRCLE DETAILS:</u> 1) Area = 1962.5cm², Circumference = 157cm.   2) 200 turns.

P.28 <u>SOLIDS AND NETS:</u>   Area of One face = 6×6=36cm², So Area of whole net = 6×36=216cm².

Answers

Section Two — Acid Tests (continued)

P.29 VOLUME AND CAPACITY: 1) 2×2×4=16.   2) 3×2×1=6cm³.
3) 8m³.   4) 96cm³.

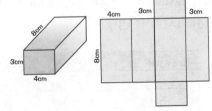

P.31 SYMMETRY:
T : 1 line of symmetry, No Rotational symmetry,
K : 1 line of symmetry, No Rotational symmetry,
I : 2 lines of symmetry, Rotⁿ symmetry Order 2,
N : 0 lines of symmetry, Rotⁿ symmetry Order 2,
S : 0 lines of symmetry, Rotⁿ symmetry Order 2,
M : 1 line of symmetry, No Rotational symmetry.

P.35 REGULAR POLYGONS: 1) See P.34   2) See P.34 3) See P.35   4) A = 45⁰   B = 135⁰

Revision Test for Section Two

1) See P.22 ; Perimeter = 32cm. 2) a) A = L × W   b) A = ½×b×hᵥ
3) A = π × r²   C = π × D.   4) See P.24.   5) a) 40cm² b) 20cm² c) 6m² d) 78.5cm² 6) 24m².
7) π is 3.14; 8m. 8) 75.4cm. 9) 452.2cm². 10) 26.5 turns. 11) See P.28. 12) a) 125cm³ b) 90cm³.
13) Line Symmetry, Plane Symmetry and Rotational Symmetry
14) See P.32/33 15) Show "Teach"! 16) See P.34.   17) 72⁰ and 108⁰.   18) Show "Teach"!

Section Three — Acid Tests

P.38 IMPERIAL AND METRIC UNITS: 1) a) 300 cm   b) 40 mm   2) a) 1.5 kg   b) 2 litres
3) 3 feet 4 inches 4) a) 100 or 110 yards b) 180cm.

P.40) ROUNDING OFF: 2) a) 2.3   b) 4.6   c) 3.3   d) 9.9   e) 0.8   3) a) 2   b) 2   c) 5   d) 1   e) 5
P.41) 1) a) 780   b) 590   c) 40   d) 30   e) 100   2) a) 3600   b) 800 c) 300
P.42 ESTIMATION/APPROXIMATION: 1) 3   2) Between 25cm² and 70cm²
3) Between 10m² and 60m².

P.43 CONVERSION GRAPHS: 1) a) 40km   b) 72km   2) a) 12½ miles   b) 31 miles.
P.45 CONVERSION FACTORS: 1) 120kg   2) 12 pints
P.46 FRACTIONS: 1) 0.6   2) £600

3) a) ²⁄₃ b) ⁵⁄₆ c) ⁵⁄₇.

P.47 FRACTION/DECIMAL/PERCENTAGE:

P.51 PERCENTAGES: 1) Type 1, £70.
2) Type 2, 9%.

Fraction	Decimal	Percentage
1/5	0.2	20%
2/5	0.4	40%
4/5	0.8	80%
1/10	0.1	10%
7/10	0.7	70%
3/8	0.375	37½%
5/8	0.625	62½%

Revision Test for Section Three

1) a) 100   b) 1000   c) 1000   d) 1000   2) 2 feet 2 inches   3) 2.2m   4) 1600cm³ 5) a) 6.4   b) 5.5   c) 8.3
6) a) 2   b) 2   c) 5   d) 11   7) a) 530   b) 500   8) a) 3000   b) 50   9) a) 2   b) 2   c) 2   d) 3   e) 1   10) 10
11) 10m.   12) 4m².   13) See P.44.   14) 375cm   15) a) £8.   b) 170 French Francs.   16) 21 miles .

17) 2590 kg.   18) Fractions. a) £17.   b) £40.   19) a) 0.125.   b) ¾ .

20) a) Divide.   b) × by 100.   21) ½ = 0.5 = 50%.   22) 1/5 = 0.2 = 20%.   23) £7 saved: cost = £21.
24) 20% reduction.

Answers

Section Four — Acid Tests

P.55 MEAN/MEDIAN/MODE/RANGE: First rearrange them: 2, 3, 5, 7, 8, 10, 10, 11, 16 (9);
mean = 8; median = 8; mode = 10; range = 14.

P.56 TALLY TABLE:

Height h (cm)	Tally	Frequency
J	l	1
K		0
L	ll	2
M	ll	2
N	llll	5
P	ll	2
R	llll	4
S	l	1
		Total 17

P.57 BAR CHART:

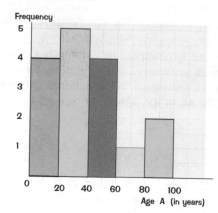

P.59 GRAPHS AND CHARTS: 2) They are closely related. They have POSITIVE correlation.

P.60 PIE CHARTS:

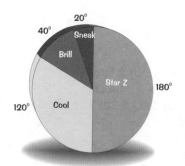

P.63 PROBABILITY: 1) 1 or certain.
2) a) 1/6.   b) 5/6.
3) 4/15.
4) H-1, H-2, H-3, H-4, H-5, H-6,
T-1, T-2, T-3, T-4, T-5, T-6.

P.64 PROBABILITY EXPERIMENTS: Show "Teach"!

P.65 FIRST QUADRANT COORDINATES:

(1,1) (4,1)
(1,4) (4,4)

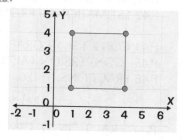

P.66 POSITIVE AND NEGATIVE COORDINATES:
1) a) A(-3,-3);   B(-3,1); C(0,3); D(3,1); E(3,-3). b) and c) Show "Teach"! d) AB is x=-3; AE is y=-3.

P.67 FOUR GRAPHS:

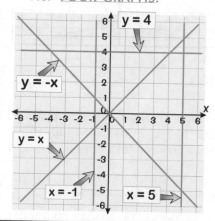

P.69 DRAWING GRAPHS:

x	-4	-2	-1	0	1	2	4
y	-1	1	2	3	4	5	7

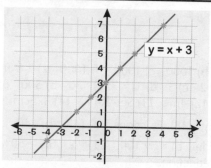

Answers

Revision Test for Section Four

1) First put in order: 2, 2, 3, 5, 6, 7, 9, 10. (8)   a) Mode=2   b) Median=5.5.   c) Mean=5.5.   d) Range=8.
2) Pictogram. 3) 35.   4) Scattergraph.   5) Quite good Negative Correlation.

6)

Fruits	Apple	Strawberry	Banana	Orange	Total
Number	20	15	10	15	60
Angle	120	90	60	90	360 degrees

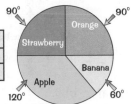

7) Values of w from 20 to 30 <u>including</u> 20 but <u>not including</u> 30.
   30 would be in the next group.
8) 9/27 or 1/3. 9) HH   HT   TH   TT; 1/4+1/4=1/2.
10) H1 H2 H3 H4 H5 H6
    T1 T2 T3 T4 T5 T6; 1/12.
11) 0.7.

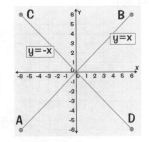

12)   Line AB is y=x;
13)   Line CD is y=-x.

Section Five — Acid Tests

P.73 <u>CLOCKS AND CALENDARS</u>:1) a) 6:30pm b) 2:45pm   2) 11:15am 3) 38.
P.74 <u>COMPASS DIRECTIONS AND BEARINGS</u>:   1)   2) 3 right angles.
   3) 180⁰.
   4) 225⁰; 315⁰.

P.75   1) a) 120⁰ ; b) 300⁰ ; 2) a) 050⁰ ; b) 180⁰ .
P.77 <u>MAPS AND SCALES</u>: 2) 400m   3) 6cm.
P.78 <u>LINES AND ANGLES</u> 1) a) 60⁰   b) 80⁰   c) 100⁰   d) 210⁰.
P.79 <u>MEASURING ANGLES</u>   3)

 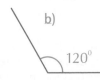 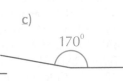

a) 55⁰   b) 120⁰   c) 170⁰

P.80 <u>FIVE ANGLE RULES</u>: 2) 55⁰ .

P.81 <u>3-LETTER NOTATION</u>: 1) 30⁰   a) BAC or CAB.   b) BAD or DAB.
P.82 <u>CONGRUENCE</u>: 1) Congruent shapes are the same size and same shape.   2) Show "Teach"!

P.83 <u>REFLECTION AND ROTATION</u>:
   4) A → B is a rotation of 180⁰ about the origin.
   5) A' → B is a reflection in the Y-axis
      A" → B is a reflection in the X-axis.

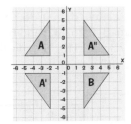

Answers

Revision Test for Section Five

1) 8:20pm. 2) 19:30. 3) 5 hours and 50 mins . 4) 2.45pm. 5) a) 31; b) 30; c) 31.
6)

7) a) 2; b) 3.
8) "FROM", NORTHLINE, CLOCKWISE.
9) and 10)

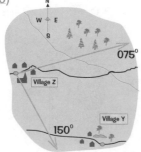

11) 305⁰.
12) a) 3m×3.5m b) 2m long, 1m wide.
13) 32km.
14) See P.78.
15) a) 37⁰ b) 80⁰ c) 162⁰ d) 288⁰
16) See P.79
17) x =115⁰, y = 50⁰
18) Congruent.
19) See P.83.

Section Six — Acid Tests

P.87 BALANCING: 1) 6 2) 3 3) 3 4) 3.
P.88 POWERS: 1) 64 2) 25; 216 3) 3 4) 162.
P.89 SQUARE ROOTS: 1) 8 2) a) 6 b) 6.325 c) 19683 d) 256 3) x = 4m.
P.91 NUMBER PATTERNS: A TYPICAL QUESTION: 1) 4, 6, 8, 10 2) a) 12 b) 14 c) 20.
P.93 NUMBER PATTERNS FORMULA: 1) a) 34, 41 Add 7. b) 8000, 80000 "Multiply the previous term by 10" c) 4, 2 "Divide the previous term by 2" 2) 2n + 3 3) a) 8. b) 2n+2.
P.95 NEGATIVE NUMBERS:1) -21, -11, -10, -4, -1, 0, 5, 6, 20, 22. 2) 13⁰C 3) a) -50 b) +3.
 4) y = +4 y = +9 5) -12
P.96 BASIC ALGEBRA: 1) a) 5x + 8 b) 7y – z 2) a) 12gh – 4g b) 35d – 14 – 28f².
P.97 TEMPERATURE ⁰F AND ⁰C: 1) 77⁰F Phew! Hot.
 2) 5⁰C Cold, but not frosty (though some risk of dense fog patches I'd say).
P.98 MAKING FORMULAS: 1) Y = 7X – 6. 2) C = 52n.
P.99 TRIAL AND IMPROVEMENT: 1) x = 3 2) x = 7.4.

Revision Test for Section Six

1) 6 2) a) 64 b) 7776 c) 64 d) 20736 e) 2000 f) 343 3) x = 7 4) a) 8 b) 7 c) 10.
 5) 7x – 5. 6) 14. 7) 9cm 8) i) 26, 32; add 6 to the previous term ii) 24, 32; add one extra each time.
iii) 162, 486; multiply by 3. iv) 110, 55; divide by two. v) 19, 15; subtract one less each time.
9) nth number = 3n + 4 . 10)

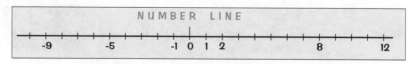

11) 11⁰C 12) a) +14 b) –16 c) –30 d) +3. 13) 3x + y – 7. 14) 10m + 15n – 20. 15) 8g + 20gh – 24gm.
16) G = 62. 17) ⁰F and ⁰C 18) 70⁰F 20⁰C 19) 95⁰F Very hot day! 20) N = 2M + 3. 21) C = 27n.
22) C = Pn. 23) a) x = 4 b) x = 2. 24) x = 7.7. 25) Z = 2.9. 26) a) 16 b) 19 c) 64.

Index

Index